# STEEL SQUARE

# BOOKS OF THE BUILDING TRADE SERIES

*Published by the American Technical Society*

BLUEPRINT READING FOR THE BUILDING TRADES
*Dalzell-McKinney-Ritow*

BUILDING INSULATION
*Close*

BUILDING TRADES BLUEPRINT READING
VOLUME I—*Dalzell*
VOLUME II—*Dalzell*

CARPENTRY
*Townsend*

CONCRETE DESIGN AND CONSTRUCTION
*Gibson*

FUNDAMENTALS OF CARPENTRY
VOLUME I—Tools, Materials, Practice—*Durbahn*
VOLUME II—Practical Construction—*Durbahn*

HOW TO DESIGN AND INSTALL PLUMBING
*Matthias, Jr.*

HOW TO ESTIMATE FOR THE BUILDING TRADES
*Townsend-Dalzell-McKinney*

HOW TO GET THE MOST HOUSE FOR YOUR MONEY
*McNamara*

HOW TO PLAN A HOUSE
*Townsend-Dalzell*

HOW TO REMODEL A HOUSE
*Dalzell-Townsend*

INTERIOR ELECTRIC WIRING AND ESTIMATING
*Uhl-Nelson-Dunlap*

MASONRY SIMPLIFIED
VOLUME I—Tools, Materials, Practice—*Dalzell-Townsend*
VOLUME II—Practical Construction—*Dalzell-Townsend*

PAINTING AND DECORATING
*Dalzell-Sabin*

STAIR BUILDING
*Townsend*

STEEL CONSTRUCTION
*Burt-Sandberg*

STEEL SQUARE
*Townsend*

**GABLE AND VALLEY ROOF**

Courtesy of Weyerhaeuser Sales Co., St. Paul, Minn.

# STEEL SQUARE

USE OF THE SCALES
ROOF FRAMING
ILLUSTRATIVE PROBLEMS
OTHER USES

GILBERT TOWNSEND, S.B.
*member of*
ROSS, PATTERSON,
TOWNSEND, AND
HEUGHAN
*Architects and Engineers*
*Montreal, Canada*

*ILLUSTRATED*

1948

AMERICAN TECHNICAL SOCIETY
CHICAGO, U.S.A.

# PREFACE TO FIRST EDITION

The steel square is the carpenter's handbook, instructor, and tool without price. Without it he is just another hammer-and-saw man— a fellow who can hit a nail and saw to a line. The ability to *make* the lines is what constitutes the difference between a real craftsman and a hammer-and-saw man. It requires technical knowledge of a high order to make most of the lines, and without a steel square a carpenter would have to be an expert mathematician and understand the mysteries of geometry. The steel square brings him a knowledge of lines and angles in a simple and practical way. To watch an expert carpenter lay out the various cuts on the rafters and other members of a complicated roof structure is a joy.

This book on the steel square tells the carpenter "how to do it." It explains clearly, by means of illustrations and detailed instructions, the various markings which appear on the square, and the purposes for which they are used by the carpenter in the course of his daily work. Instruction is also given on the use of the square in the construction of the different parts of wood-framed buildings, including detailed information on the building of roofs. The laying out of the various cuts is illustrated in a step-by-step fashion which makes some of the most difficult operations seem easy.

We feel sure that the book will prove an invaluable aid to the young carpenter who may have to rely entirely on what knowledge he can pick up in working as a helper with a man who knows. The expert carpenter is not always a good teacher; and therefore we are sure that, as a result of his study of this book, the young worker will be able to benefit to a greater degree from his daily experience. We also feel sure that carpenters of many years' experience will find new short-cuts and helps for doing jobs which are out of the usual run of their work. It is our hope that the book will prove to be another fine tool which can be added to the kit of everyone who is starting out with a bright new steel square in his tool chest and ambition to become an expert in its use and a master of his craft.

# PREFACE TO SECOND EDITION

The constant companion of every skilled carpenter is the steel square. The new edition of *Steel Square*, which has been enlarged and improved by the addition of an "Illustrative Problem," proves this fact.

All carpenters have a general knowledge of the steel square. Many, however, desire to be shown exactly and in detail how this tool may be applied to find the answer to questions which arise in the building of a house.

With this in mind, the author has followed the construction of a typical house from start to finish. The many problems confronting the carpenter are explained. Then, step by step, the solution of these problems with the use of the steel square is given. The difficult work of cutting and fitting the timbers is dealt with, just as it is met and handled in the actual building of the frame of this particular house. In every phase of the construction, the way in which the steel square can and should be used is carefully explained. Application of this tool to the job of marking off the necessary cuts on the material is shown by many clear illustrations. The inexperienced carpenter, as well as the experienced carpenter, will find in this "Illustrative Problem" the answer to many puzzling questions. From the laying out of the wall lines to the construction of the dormer, the steel square is a valuable aid.

This second edition of *Steel Square* combines in a unique manner the fundamental knowledge of the use of the square and its actual use on the job. In no other book which we have seen has the use of the steel square been explained in just this way. It is our hope that the novel addition of the "Illustrative Problem" in the book will add to its usefulness and make it an even more welcome addition to the carpenter's tool kit or to the library of a vocational school.

*The text material in this book is included in the cyclopedia "Building, Estimating, and Contracting."*

# CONTENTS

HIP AND VALLEY ROOF WITH GABLE
Courtesy of Dant & Russell, Inc., Portland, Oregon

# THE STEEL SQUARE

## CHAPTER I

## INTRODUCTION

It has been said that one of the principal reasons for the development of man through the ages to the point where he became able to make use of the forces of nature for his own benefit is the ability, which he and only he among living things possesses, of inventing, making and using tools. Probably the first tools which men made and used were weapons to aid them in hunting and fighting, but the next progressive step must have been to build rude shelters of different sorts, depending upon the climate of the countries where they lived, and this must have led to the development of building tools. Thus these, together with rude farming implements, were the earliest tools used. Ancient drawings illustrate some of them and specimens of others may be seen in the great museums.

Among these tools, along with the hatchets, chisels, hammers, saws, etc., is to be seen the *square*. Its use in those ancient times was, as it is today, to insure that the different parts of the work are *true* and *square* and therefore will fit together properly. These early squares were of course not steel squares, since steel was then unknown. All the tools of those times seem to have been of wood or stone or bronze, but it is evident that the use of some sort of square by some sort of carpenter has been common practice for thousands of years.

With the passage of time, the square became one of the most useful tools in the carpenter's kit, and especially so for the more advanced workman who laid out the work for others less clever or less ambitious than he. Because it necessarily had straight, true edges, the square came to be used as a measuring tool, and these edges came to be marked off in inches and subdivisions of inches, such as halves, quarters, eighths, sixteenths, and twelfths. Because the square had broad, smooth sides, was in early times made of wood, and was always handy, it probably became the custom among the lead-ing tradesmen to mark down on these smooth surfaces certain rules, notes, and tables, which they discovered and passed on to each other even before the use of books and note paper became common.

With the advent of steel, it became possible to manufacture squares which would remain permanently absolutely true and on whose surfaces and edges could be etched a variety of lines, scales, and tables, which gradually became standardized. New uses for these markings and for the square were discovered from time to time, until the steel square became for the intelligent carpenter a calculating machine of most remarkable usefulness. It became for him what

Fig. 1. Steel Square in Use

the slide rule is for the Engineer, what handbooks of rules and tables are for the office-man, and what the T-square is for the draftsman.

In spite of the wide variety of its uses and in spite of the fact that its many markings seem bewildering to one not accustomed to it, the steel square is easy to understand after its various uses have been explained.

## GENERAL DESCRIPTION

Modern squares, which can now be bought in any hardware store, are made of steel in various finishes, such as polished, blued, galvanized, copper-finished, and nickel-plated, or they may be had in stainless steel. The copper-finished and blued squares have the figures and graduations picked out in white enamel.

**Parts of a Steel Square.** Fig. 1 shows the shape of the tool and illustrates in a general way how it is used. It will be seen that the square is comparatively thin, yet it is thick enough so that it is not

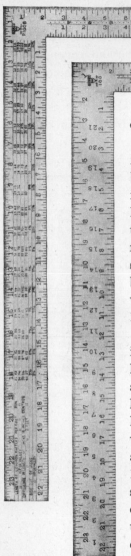

Fig. 2. Faces of Two Typical Steel Squares
*Courtesy of The Stanley Rule and Level Plant, New Britain, Conn.*

flexible. Fig. 2 shows that it has two arms or blades which meet at right angles, one of which is longer and broader than the other. The longer, wider part is called the *body* or is sometimes called the *blade*, and the shorter, narrower part is called the *tongue*. In the standard steel square which is in most common use, the blade, or body, is 24 inches long and 2 inches wide, while the tongue is 16 inches long and 1½ inches wide.

Although probably nine out of ten squares in use are of the size mentioned, it is well to know that some are slightly different. For example, the tongue is sometimes made 18 inches long and 1½ inches wide instead of 16 inches long, and is even made as short as 12 inches; while some of the older squares made years ago, but which may still be in use, have tongues 14 inches long by 1½ inches wide. Special squares are also made with the blade 18 inches or 24 inches long by 1½ inches wide and the tongue 12 inches long by 1 inch wide, and other small ones with the blade 12 inches by 1½ inches and the tongue 8 inches by 1 inch.

Fig. 2 illustrates two different squares, the one marked *R100B* having more markings on it than the other one and consequently being more expensive. The different parts of the steel square are illustrated in Figs. 3 and 4.

The outside corner, where the two outside edges of the blade and the tongue of a square meet at right angles, is called the *heel* of the square.

The inside corner at which the two inside edges meet is also sometimes referred to as the *heel*.

One side of the square is called the *face* and the other side is called the *back*. The face can be recognized easily because the name of the manufacturer is always to be found stamped on this side.

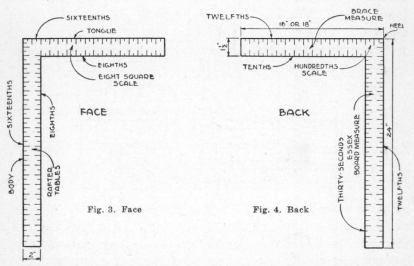

Figs. 3 and 4. Location of Scales and Tables on Face and Back of a Steel Square

(See Fig. 2.) Another way to find out which side is the face is to hold the square with the body or blade (the longer, wider part) in the left hand and the tongue (the shorter, narrower part) in the right hand with the tongue pointing *toward* you, see Fig. 1. When holding the square in this way, you will be looking at the face; therefore, by turning the square over and looking at the opposite side, you will be looking at the back.

**The Take=Down Square.** A square called a *take-down square* is also manufactured and sold. It can be taken apart at the joint between the blade and the tongue and packed away in a canvas case (which is sold with it) so that it can be carried about more conveniently. This square has a cam locking device, so that by turning the cam with a screwdriver the tongue is drawn firmly into place. The parts are so made that any possible wear will be taken care of automatically and the square will always be correct when the two parts have been locked together.

**Testing a New Steel Square.** One of the most important things about a steel square is that it must be truly square; that is, the two arms, namely the blade and the tongue, must be truly and exactly at right angles.

When using a new or untried square for important work, it is sometimes desirable to test it to make sure that it has been made exactly right. The way to satisfy yourself on this point is as follows:

Smooth up one side of a very wide board which is at least 4 feet long, and true up one edge until it is perfectly straight. Lay this

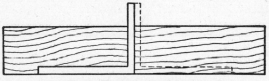

Fig. 5. Testing a Steel Square

prepared board down with the straightened edge nearest to you and the smoothed side up. Lay the square flat on top of the board as shown in Fig. 5, with the outside edge of the blade extending to the left and lying exactly on top of the trued edge of the board throughout the entire length of the blade. The tongue will then be extending away from you across the board. Now, take a pen-knife or a very sharp hard pencil and, while holding the edge of the blade always truly in line with the straight edge of the board, make a mark on the smooth face of the board close against the edge of the tongue of the square. Next, turn the square over, keeping the heel of the square at exactly the same point, but so that the blade will in this case extend to the right along the straight edge of the board and exactly in line with this edge throughout the entire length of the blade. Now, holding the edge of the blade of the square always truly in line with the straight edge of the board and keeping the heel of the square exactly where it was before the square was turned over, compare the position of the outside edge of the tongue in its new position with the mark which you previously made across the board. If the edge of the tongue is still exactly on the mark, or if a new mark made with the pen-knife blade or pencil against the edge of the tongue in its new position will come exactly on top of the first mark, then the square is truly *square*.

## SCALES

As has already been said, the edges of the earliest squares, used ages ago, were probably used as rules, because they were straight and true. For this reason, these edges came to be marked off in inches and parts of an inch like a rule. On some of the older tools these markings were different on different squares, but they have now come to be well standardized, so that the same kinds of markings, called *scales*, are to be found in the same place on all steel squares. The scales and their location on the squares will now be described.

If you take the longer and wider part of the square (the body or blade) in your left hand, and the shorter and narrower part (the tongue) in your right hand, with the tongue pointing toward you, you will be looking at the face. Now all around the outside edge on both the blade and the tongue you will find marks indicating inches and sixteenths of an inch and all around the inside edge on both the blade and the tongue you will find inches and eighths of an inch. This is shown on Fig. 3. The scale starts at the heel of the square in all cases.

Now, turn the square over or take it with the blade in the right hand and the tongue in the left hand with the tongue pointing toward you, and you will be looking at the back of the square. All around the outside edge on both the blade and the tongue you will find inches and twelfths of an inch, see Fig. 4. On the inside edge of the tongue you will find inches and tenths of an inch (a scale that is little used), and on the inside edge of the blade you will find inches and thirty-seconds of an inch. This is true of the more expensive squares. On some of the cheaper squares the division of the inches into thirty-seconds of an inch is omitted and this edge is divided only into eighths. Likewise, on some cheaper squares the division of the inches into tenths of an inch is omitted, and this edge is divided only into eighths. Also, on some smaller or cheaper squares, the scales along the edges are divided only into quarters and eighths of an inch, the division into sixteenths being omitted.

**Hundredths Scale.** On some of the more expensive squares, still another scale is to be found, called the *hundredths scale*. This is not on any of the edges of the squares, but is always located on the back of the tongue near the heel as shown in Fig. 6. The hundredth scale is a scale one inch long, which is divided into exactly one hundred equal parts. These subdivisions are necessarily very small,

Fig. 6. Showing the Hundredths Scale near the Heel
*Courtesy of The Stanley Rule and Level Plant, New Britain, Conn.*

but a workman with good eyes and with the aid of a pair of angle dividers can take off any required number of hundredths of an inch.

This introduces a tool which is used frequently in connection with the steel square, and as reference will be made to it again in the following pages, it will be described here. Fig. 7 illustrates the dividers. They can be used with or without the pencil attachment shown, and can be adjusted to measure off any distance within their range and then clamped in place by means of the setscrew.

In addition to the hundredths scale, the two sides of the steel

Fig. 7. Dividers and Pencil Clasp
*Courtesy of The Stanley Rule and Level Plant, New Britain, Conn.*

square are covered with other scales and tables, which will be described briefly. The use of these scales and tables will be explained in detail further on.

**Octagon Scale.** If you take the square with the body or blade (the longer, wider part) in the left hand and the tongue in the right hand, with the tongue pointing toward you, you are looking at the face of the square. Now along the middle of the tongue (the shorter, narrower part) you will see a scale which is called the *octagon scale* and which is also sometimes called the *eight square*. This scale is used for obtaining the necessary measurements to enable the workman to shape a square timber into a piece having eight sides, or, in other words, an *octagon*. The use of this scale for this purpose will be explained in detail later on. The cheaper squares do not have this scale. It is shown in Fig. 2.

**Essex Board Measure.** In the middle of the back of the blade a series of figures known as the *Essex board measure* will be seen. It is used for obtaining quickly and easily the contents in "feet board measure" of any size timber. Feet board measure means the number of square feet of lumber one inch thick. A piece 12 inches wide and 1 foot long and 1 inch thick contains one foot board measure or F.B.M. If the piece is 10 feet long instead of 1 foot long, it will contain 10 F.B.M.; if it is 2 inches thick, it contains twice as many F.B.M.; if it is 3 inches thick, it contains three times as many F.B.M. as a piece 1 inch thick, or 30 F.B.M.; and so on. If the piece is only 6 inches wide instead of 12 inches wide, it will contain half as many F.B.M.; if 9 inches wide, it will contain 9/12 or 3/4 as many F.B.M.; and so on. This is the way in which lumber is measured and bought and sold. You buy so many thousand F.B.M. and you pay so much per thousand F.B.M., or so many dollars a thousand.

*Use of the Steel Square in the Calculation of Board Measure.* You can find the contents in feet board measure for any size timber by arithmetic if you know how, but if you do not, or if you do not want to take the time, you can find the F.B.M. directly from the back of the blade of your steel square. Holding the blade in your right hand and the tongue in the left hand, with the tongue pointing toward you so that you are looking at the back of the blade, Fig. 8, you will notice along the outside edge, beginning at the left at the heel, the regular inch divisions, *1, 2, 3, 4, 5,* etc. In using the square to find feet board measure, these figures are used to show the width in inches of the stick of timber which is being considered, such as 8 inches wide, 9 inches wide, 12 inches wide, etc. Under each of these widths will be found seven other figures which give directly in feet (to the left of the vertical line) and in twelfths of a foot (to the right of the vertical line) the F.B.M. in timbers of that width one inch thick and of seven different lengths. The seven different lengths, starting

at the top, are: 8 feet, 9 feet, 10 feet, 11 feet, 13 feet, 14 feet and 15 feet, and you will notice that these figures appear under the 12-inch mark, as shown in Fig. 8.

Thus you have only to find under the 12-inch mark the figure corresponding to the length (in feet) of your stick of timber and follow back to the left along the horizontal line until you come to the point under the mark corresponding to the width (in inches) of your stick, and there you will find a figure which is the contents of the stick of timber in feet board measure. The figure appearing at the left-hand side of the vertical line is full F.B.M. and the figure at the right of the vertical line is twelfths of an F.B.M.

*Example 1.* Find the F.B.M. in a board 1 inch thick, 10 feet long and 9 inches wide.

Begin by looking in the column of figures underneath the 12-inch mark on the outside edge of the back of the blade and near the middle of the blade you will find the figure 10. Follow along to the left on the horizontal line underneath the figure 10 until you come to the column of figures underneath the 9-inch mark and there you will find the figures 7/6, which stand for seven and six twelfths feet board measure, which is the F.B.M. of the stick of timber. If the piece of lumber were thicker than 1 inch, you would multiply this result by the thickness of the piece in inches. If the piece were more than 12 inches wide, you would follow the horizontal line underneath the

Fig. 8. Back of Blade of Steel Square
Showing Essex Board Measure

figure 10 in the 12-inch column to the right instead of to the left.

A length of 15 feet is the longest indicated in the column of figures underneath the 12-inch mark on the outer edge of the back of the blade. If the F.B.M. is required for a stick longer than this, it can be found by following the rule given above in *Example 1* but using only half the actual length and then doubling the result, since it is clear that doubling the length of a piece of timber doubles the contents in F.B.M.

In order to show how to deal with a larger and longer stick of timber, another example is given.

*Example 2.* Find the F.B.M. in a timber 10 inches wide, 16 inches deep and 23 feet long.

Divide the length of 23 feet into two parts, 10 feet and 13 feet. Let the 10-inch dimension be taken as the width and consider the timber to be made up of 16 separate boards each 1 inch thick and 10 inches wide. Find the F.B.M. for one of these boards and multiply the result by 16 to get the whole F.B.M. of the piece of lumber. Following the procedure used in the previous example and referring to Fig. 8, a 1-inch board 10 feet long and 10 inches wide contains $8\frac{4}{12}$ F.B.M. and a board 13 feet long and 10 inches wide contains $10\frac{10}{12}$ F.B.M. Adding these together gives $19\frac{2}{12}$ F.B.M. and multiplying by 16 gives $306\frac{8}{12}$ F.B.M. as the contents of the entire stick.

# CHAPTER II

## HOW TO USE THE SQUARE—BRACES

**Making a Fence.** Another very useful tool to be used along with the steel square is a wooden instrument which can easily be made by any carpenter and which is called a *fence*.

To make one, take a piece of cherry or walnut or similar hard wood dressed down to a width of 2 inches and a thickness of 1⅝ inches and of any convenient length a little less than 4 feet. Mark off a line all along the exact center of each edge and using this mark as a guide, make a cut (called a *kerf*) with a saw, extending at least 18 inches from each end toward the middle, leaving about

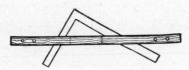

Fig. 9. A Fence for Use with a Steel Square

8 inches of the piece at the center uncut. The kerfs must be wide enough so that the thickness of the steel square can pass into them as shown in Fig. 9. See also Fig. 83.

Now put the fence on the steel square, so that the blade of the square passes into one kerf, and the tongue of the square into the other. The fence thus forms the third and longest side of a triangle, with the blade and the tongue of the steel square forming the other two sides. To make the fence stay in any desired position on the square, put one or more No. 10 size 1½-inch screws in each end so that the saw cut or kerf may be tightened up at will to hold tight against the steel square.

The object of the fence is to provide a means of placing the square repeatedly in the same relative position along a piece of lumber as shown in Fig. 10 and of doing it quickly and easily and at the same time accurately.

**Use of the Steel Square in Making a Brace.** To illustrate the use of the fence and to show how the square can be used in solving a comparatively simple problem, explanation is given of how to make a *brace* such as is shown in Fig. 11. In this figure the upright member *A* might be the corner post in a braced house frame and the horizontal member *B* might be a girt at the second floor level. The member *C* is the brace which is framed into the other two members. The bottom of the girt *B* intersects the inside edge of the upright *A* at the point *e*. The outside edge of the brace *C* intersects the inside edge of the corner post *A* at the point *f* and it inter-

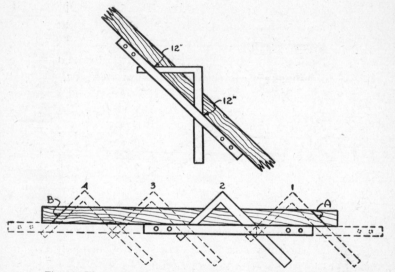

Fig. 10. Illustrating the Use of the Square and Fence in Making a Brace

sects the bottom of the girt *B* at the point *g*. The distance along the girt is called the *run* of the brace horizontally and the distance *ef* along the corner post is the *vertical run*. Suppose that these two runs are equal, each being 4 feet or 48 inches. The length of the brace will be about 68 inches or 5⅔ feet. How to find this length by means of the steel square will be explained later.

Take a piece of scantling of the required size, say 4 by 4 inches and dress one edge straight and true. Take the steel square with the fence loosely fitted to it, as shown in Fig. 9, and apply it to the piece of scantling as shown in Fig. 10, and move the square around until the 12-inch marks on the outside edge of both the blade and the tongue are exactly in line with the dressed edge of the piece of scantling. Holding the square in this position securely, so that it will not slip, move the fence up along the blade and tongue until it comes against the dressed edge of the piece of scantling and rests snugly against it. Make sure that the 12-inch marks on the blade and tongue of the square still coincide exactly with the edge of the piece of scantling and then clamp the fence tight on the square by means of the screws. Of course, the edge of the fence will also cross the outside edges of the blade and tongue of the square exactly

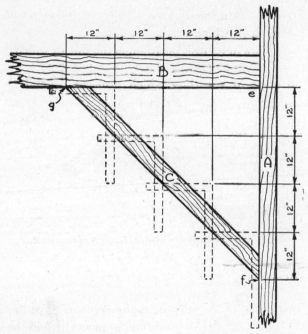

Fig. 11. Brace in Place Showing Four Positions of Square

on the 12-inch marks. Be sure the measurement is exact, because even a small error or a little carelessness at this point may lead to a large error later on, as will be seen.

Having prepared the square with its fence for use in this way, slide it along nearly to the right-hand end of the piece of scantling, as shown by the dotted outline of the square at *1*, and make a mark along the outside edge of the blade and another one along the outside edge of the tongue where it crosses the dressed edge of the piece of scantling. Next, move the square along to the left until the outside edge of the blade crosses the dressed edge of the piece of scantling at the same point at which the outside edge of the tongue crossed it before. Now make a mark along the outside edge of the tongue of the square where it crosses the dressed edge of the piece of scantling in its new position and again move the square along to the left the same as before until the outside edge of the blade crosses the dressed edge at this new mark. Slide the square along to the left in the same way once more.

Note that the square was placed on the piece of scantling four times, that is, just as many times as there are feet in the run of the brace. The mark A, which was made along the outside edge of the blade of the square in its first position, is the line in which the lower end of the brace will fit against the upright corner post. The small end of the piece of scantling which projects beyond this mark may be formed into a tenon to fit into a mortise to be cut in the post, or the brace may be cut off along this mark and simply spiked against the post. When the square is in its fourth and last position on the piece of scantling, if a mark is made all along the outside edge of the tongue at B, this mark will be the line in which the upper edge of the brace will fit against the under side of the girt. The small part of the piece of scantling forming the brace which projects beyond this mark may be formed into a tenon to fit into a mortise to be cut in the under side of the girt, or the brace may be cut off along this mark and simply spiked against the girt.

Fig. 10 shows the brace in a horizontal position as it would be when being measured and laid out, and shows the four positions of the steel square on it, one position being shown by full line and the three other positions by means of dotted lines.

Fig. 11 shows the same brace placed between the corner post and the girt. The four positions of the square are indicated on it by means of dotted lines just as in Fig. 10. Note that every time the square was moved along the piece of scantling to a new position, with the 12-inch marks on the outside edge of the square kept always in line with the dressed edge of the scantling (the outside edge of the brace), the square was advanced 12 inches in the direction of the run of the brace.

The procedure is the same for a brace of any length, provided that the vertical and horizontal runs are both the same, the only difference being that whereas for a 4-foot (or 48-inch) run the square was applied to the piece of scantling four times, for a 5-foot (or 60-inch) run it would be applied five times, and for a 6-foot (or 72-inch) run, six times, etc.

Now suppose that the run of the brace were 3 feet 6 inches, or 42 inches, both vertically and horizontally, that is both along the upright corner post and along the horizontal girt. In this case the piece of scantling would be prepared just as before, and the square would be placed with its fence crossing the 12-inch mark on the outside edges of the blade and tongue, as shown in Fig. 10. The square

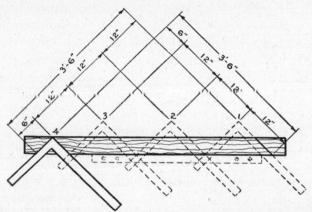

Fig. 12. Application of Square and Fence to a Brace with Run
of 3'–6"

would be placed on the piece of scantling three times in three different positions as shown by the dotted lines in Fig. 12. For the fourth position the square would be applied to the piece of scantling as shown in Fig. 12 at 4, with the 6-inch marks in line with the dressed edge instead of the 12-inch marks as for the other three positions. It will be seen from the figure that the square was moved along the piece of scantling *in the direction of the run of the brace,* a distance of 3 feet 6 inches, or 42 inches.

The question now arises, How long is the brace itself when the vertical and horizontal runs are both 4 feet (48 inches) or 3 feet 6 inches (42 inches) or any other length?

This question can be answered at once with the help of the steel square as follows:

**Brace Measure.** If you will take the square with the blade

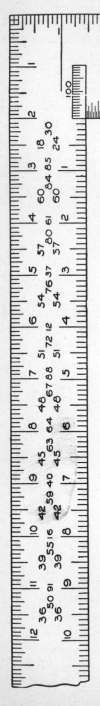

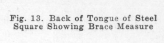

Fig. 13. Back of Tongue of Steel
Square Showing Brace Measure

(the longer, wider part) in your right hand and the tongue (the shorter, narrower part) in your left hand with the tongue pointing *toward* you, you will be looking at the back of the square, Fig. 13. On the back of the tongue, you will see along the middle a series of figures like this

$$\frac{36}{36} \; 50.91 \quad \frac{39}{39} \; 55.16$$

starting at the left-hand end with $\frac{24}{24}$ 33.94 and continuing to the right-hand end where $\frac{60}{60}$ 84.85 will be found. These figures mean that if the run along the upright corner post and along the horizontal girt were equal and were both 24 inches, then the length of the brace *fg* in Fig. 11 would be 33.94 inches; if the run were 36 inches, the length of the brace *fg* would be 50.91 inches; run 39 inches, length 55.16 inches; run 60 inches, length 84.85 inches; and so on for the other runs from 24 inches, with a difference of 3 inches more each time, up to 60 inches. This is called the *brace measure* or sometimes the *brace rule*. The cheapest squares do not have it, but most squares do.

Now you will notice that just as 60 is twice 30, so the length opposite the $\frac{60}{60}$ mark, namely 84.85, is just twice the length opposite the $\frac{30}{30}$ mark, which is 42.42. This shows that if you have a brace whose run is longer than any run shown in the brace measure, you can get its length by finding one half the actual run in the brace measure and then doubling the length given for it. Thus for a run of 78 inches, you would find $\frac{39}{39}$ 55.16 on the square and the length of the brace would be twice 55.16, which would be 110.32 inches. Looking at it the other way about, if the actual run of

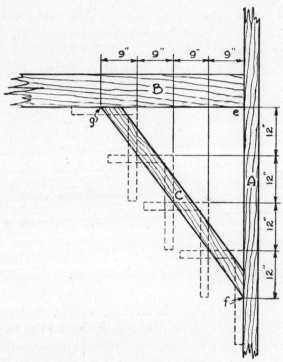

Fig. 14. Brace with Unequal Runs

your brace were 20 inches, you could not find 20 inches or 40 inches on the brace measure, but you could find $\frac{60}{60}$ 84.85, the 60 being just three times your actual run of 20 inches, and the length of your brace would be just one third of the 84.85, which is 28.28 inches.

Thus far only braces having equal vertical and horizontal runs have been considered. Of course, the vertical and horizontal runs may be different. If you will look at the extreme right-hand part of the brace measure on the tongue near the heel of the square, you will see the figures $\frac{18}{24}$ 30.  (Fig. 13.)  This means that if either run

of a brace is 18 inches and the other run is 24 inches, then the length of the brace is 30 inches.  Applying what you learned in the preceding paragraph, if one run is twice 18 inches, or 36 inches (3 feet) and the other run is twice 24 inches or 48 inches (4 feet), then the length of the brace will be twice 30 inches or 60 inches (5 feet).

Suppose that you are to lay out a large number of braces with a vertical run of 4 feet and a horizontal run of 3 feet, as shown in Fig. 14, formed of 4x4 scantlings.  The work would not be laid out for each one on the actual scantling as was done for the brace shown in Fig. 10, but a *pattern* would be made for use as a guide by which to lay out and cut all of the braces from the 4x4 stuff.  To make the pattern, select a piece of good stuff 4 inches wide, 1 inch thick, and

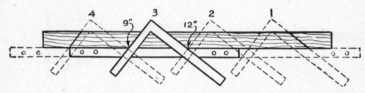

Fig. 15.  Use of Steel Square to Lay Out a Brace When Horizontal and Vertical Runs Are Unequal

a little more than 5 feet long, since, as was found from the brace measure, the length of the brace would be 5 feet.  Dress one edge straight and true.  Adjust the *fence* on the steel square so that the edge of the fence crosses the 12-inch mark on the outside edge of the blade and the 9-inch mark on the outside edge of the tongue as shown in Fig. 15, and then apply the square to the piece of stuff near the right-hand end as before and as shown in Fig. 15.  Next, move the square to the left along the edge of the piece of stuff three times, marking along the outside edge of the blade and tongue each time the same as before, so that the square will have been applied to the piece of stuff four times in all.  The horizontal run marked off on the brace pattern will then be 4x9 inches, which equals 3 feet, and the vertical run will be 4x12 inches, which equals 4 feet, as shown in Fig. 14.  The pattern can be used repeatedly to mark off and cut the braces.

# CHAPTER III

## ROOF FRAMING

Aside from stair building one of the most difficult problems which the carpenter encounters is the framing of roofs. Without the assistance of the steel square, this would be a hard problem indeed for the average craftsman, as it would involve a lot of calculation and measurement and would require a skill in mathematics not possessed by many. It is in connection with this part of the carpenter's work that he finds the steel square most useful.

**Types of Roofs.** The roofs which he is most likely to be called

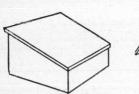 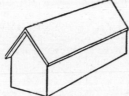 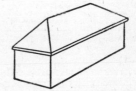

Fig. 16. Shed or Lean-to Roof     Fig. 17. Gable or Pitch Roof     Fig. 18. Hip Roof

upon to frame belong to three or four different types. Fig. 16 shows the *shed roof*, also sometimes known as the *lean-to roof*. It slopes only in one direction and is therefore the simplest type, with the exception of the flat roof.

The *gable roof*, also called the *pitch roof*, shown in Fig. 17 is the kind of roof most frequently seen. The two sloping surfaces rise upward from the side walls and meet in a line running lengthwise of the building over the center which is called the *ridge*. At each end of the building these sloping surfaces form three-sided or triangular spaces in the upper part of the end walls, which are called the *gables*. Next to the shed roof, the gable roof is the easiest kind of roof to frame.

Fig. 18 shows a *hip roof*, which is like the gable roof except at the ends of the building where there is a roof surface sloping upward from the end walls **to** the ridges so that there is **no** gable, but there are four sloping surfaces. From each corner of the building heavy rafters slope diagonally upward to meet the ends of the ridge and are framed into it, thus forming a self-supporting framework of five members, the four heavy rafters and the ridge. The ridge is strengthened by the ordinary rafters, the tops of which are framed against it. Other rafters are framed into the heavy diagonal rafters.

Fig. 19 shows a somewhat more complicated roof formed when there is an ell on the building and the two gable roofs intersect each other. This is called a gable and valley roof and the sloping line

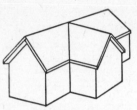

Fig. 19. Gable and Valley
Roof

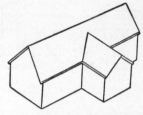

Fig. 20. Gable and Valley Roof
on a House Having a Narrow
Ell

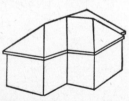

Fig. 21. Hip and Valley
Roof

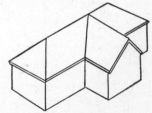

Fig. 22. Hip and Valley Roof with
Gable on Ell

in which the two intersecting roofs meet is called a *valley*. The valley slopes upward from the eaves to the ridge in the case shown in Fig. 19, but if the ell is narrow, the two valleys, one on each side of the ell, meet each other before they get up as high as the ridge. If this is the case, the ridge of the ell roof butts into the sloping side of the main roof, as shown in Fig. 20.

When the plan of the building is such that there is a projection or ell to be roofed over, both the main roof and the ell may be covered with a hip roof, as shown in Fig. 21 with a valley at the intersection similar to that in the gable and valley roof. This is called a hip and valley roof. There may be a hip roof on the main roof and a gable roof on the ell, as shown in Fig. 22, or a gable roof on the main roof and a hip roof on the ell. Both of these would be called hip and valley roofs, because they have both hips and valleys.

**Roof Framing.** Fig. 23 illustrates the framework for a hip and valley roof covering a building with an ell, having a hip roof on the

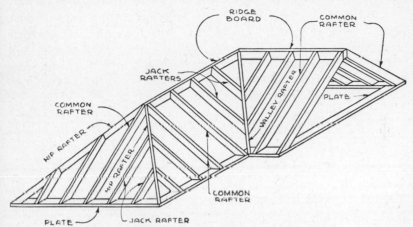

Fig. 23. Roof Frame Showing Different Kinds of Rafters

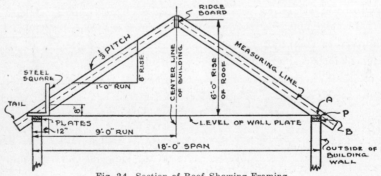

Fig. 24. Section of Roof Showing Framing

main roof and a gable roof on the ell, such as is shown in Fig. 22, but with a much flatter slope.

Fig. 24 illustrates a section taken through a roof showing the framing. If you could take the building shown in Fig. 17 and cut it in two crosswise with a huge knife and then look directly at the sliced end, you would see, as far as the framing is concerned, something like what is shown in Fig. 24. This figure only shows the part from the ridge down to the level of the top of the side walls. At the top of these two side walls are horizontal pieces called the *plates*.

The distance from outside to outside of the plates right across the building is called the *span* of the roof. If a plumb line could be dropped straight down through the center of the ridge, it would be the center line of the building and the shortest horizontal distance from this line to the outside of the plate or of the building wall would be called the *run* of the roof or the run of the roof rafters. The ridge is usually at the center of the building (though not always so) and when it is in the center, the roof is said to be equally pitched. For such roofs the run is equal to exactly half of the span or half of the width of the building frame.

Notice that the top of the rafter in each case is higher than the top of the wall plate. If a line is drawn on the side of the rafter from the outside upper corner of the plate and running upward parallel to the top edge of the rafter, this line (dotted in Fig. 24) is called the *measuring line* and is used frequently in roof framing. The two measuring lines meet at a point in the center of the ridge board somewhat below the level of the ridge. The vertical distance from this point where the two measuring lines meet, down to the level of the tops of the wall plates is called the *rise* of the roof.

The rise of a roof divided by the span of the same roof is called the *pitch* of the roof. Carpenters speak of a roof as being one-half pitch or one-quarter pitch or one-third pitch, meaning that the rise is one half the span or one-quarter of the span or one-third of the span, as the case may be. The pitch of a roof whose rise is 12 feet and whose span is 24 feet will be 12 divided by 24, which is ½ (one-half). If the rise were 6 feet instead of 12, the pitch would be 6 divided by 24, which is ¼ (one-quarter). If the rise is 6 feet and the span is 18 feet the pitch will be 6 divided by 18, which is ⅓ (one-third) as shown in Fig. 24. In all these roofs the rise is not more than one half of the span or in other words, the rise is not more than the run. This is not always the case, but it is usually so. Whatever the run (½ of the span) may be, it can be said to be a certain number of feet and if the rise is measured in inches and divided by the number of feet in the run, the result is the rise for each foot of the run. Thus in Fig. 24 the rise is 6 feet or 72 inches, and this divided by the run of 9 feet gives 8 inches rise to each foot of run.

The pitch of a roof, therefore, may be described as so many

inches of rise for each foot, or per foot, of run. A roof with a 16 foot span, 8-foot run, and 4-foot or 48-inch rise would have a rise of 48 divided by 8, which is 6 inches per foot. It will be seen that this is the same as saying that this roof has a one-quarter pitch, since the rise is one-quarter of the span.

It is useful to be able to find the rise in inches per foot of run corresponding to any pitch expressed as a fraction such as ½ pitch, ¼ pitch, or ⅓ pitch. In Fig. 25 is shown a diagram in which the figures on the steel square are used to illustrate the relation between these two methods of describing the pitch of a roof.

The distance from the heel of the square to the 12-inch mark,

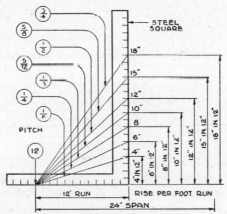

Fig. 25. Principal Roof Pitches

which is 12 inches, is taken as the run corresponding to a span of 24 inches. The rise is given as so many inches in 12 inches (inches per foot run) and the pitch corresponding to each is shown at the left. Thus a rise per foot run of 8 inches in 12 inches is shown to be the same as a ⅓ pitch, a rise per foot run of 10 inches in 12 inches is shown to be the same as a ⁵⁄₁₂ pitch, a rise per foot run of 18 inches in 12 inches is shown to be the same as a ¾ pitch (18 inches being ¾ of the span of 24 inches) and so on for the other rises and pitches.

It will be noted that if 12 inches is taken on the tongue of the square and the pitch, expressed in inches per foot of run, is taken on the blade of the square, then the line joining these two points has the same slope with relation to the tongue of the square as the rafter will have with relation to the line of the wall plates shown

in Fig. 24. In this figure, a small steel square is shown just above
the wall plate at the left-hand side, set with the tongue horizontal
and the blade upright just as it is in Fig. 25, so that the application
of Fig. 25 to an actual roof frame may be seen. The tongue of the
square lies along the line representing the level of the wall plate.
The 12-inch mark on the tongue of the square is placed at the outside edge of the wall plate, which is the outside of the building wall,
and it will be seen that this 12-inch distance represents just one foot
of the run of the roof or of the rafter. As this roof is ⅓ pitch, the
dotted measuring line will cross the outside edge of the upright

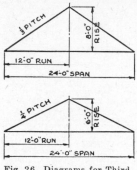

Fig. 26. Diagrams for Third
and Fourth Pitch

blade of the square at the 8-inch mark, showing that the pitch of this
roof is eight inches to the foot, and this is also shown by the triangle giving the run and rise shown just to the right of the square
in Fig. 24.

To find the rise of the roof which is the vertical distance from
the level of the wall plate up to the measuring line at the ridge,
multiply the rise, or pitch, of 8 inches to the foot by the number of
feet in the run, namely, 9 feet and the result will be 72 inches or 6
feet which is given as the rise of roof in Fig. 24. If the roof were
steeper, the measuring line would cross the outside edge of the
blade of the square above the 8-inch mark, perhaps at the 10-inch
or 12-inch or 14-inch or 16-inch or 18-inch mark as illustrated in
Fig. 25, and the pitch of the roof would vary accordingly, being
respectively 10, 12, 14, 16, or 18 inches to the foot, which is the
same as saying that the pitch would be respectively ⁵⁄₁₂, ½, ⁷⁄₁₂, ⅔, or
¾. Fig. 26 shows run and rise for one-third and one-fourth pitch.

Sometimes the pitch of a roof is spoken of as being 30 degrees or 45 degrees or some other number of degrees, meaning the angle which the slope of the dotted measuring line (Fig. 24) makes with the horizontal level of the wall plate. Thus ½ pitch, which is 12 inches in 12 inches, is a 45-degree pitch. In Fig. 27 is shown a diagram for finding pitches of various degrees on the square. Thus if the pitch is 45 degrees, use 12 on the tongue and 12 on the blade; if the pitch is 30 degrees, use 12 on the tongue and 6¹⁵⁄₁₆ on the blade. It

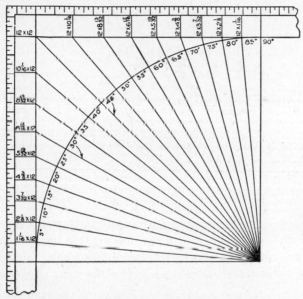

Fig. 27. Diagram for Finding Pitches of Various Degrees by Means of the Steel Square

will be seen that the 30-degree pitch is a little steeper than a ¼ pitch which is 6 on the blade and 12 on the tongue (see Fig. 25).

Instead of having a ridge along the center, the roof of a building sometimes has a flat space at the top. This flat space is called a *deck* and it may be of any width. A section through such a deck roof is shown in Fig. 28. To get the run of the rafters in this roof it is necessary to subtract the width of the deck from the span and divide the remainder by two. The result will be the run. In Fig. 28, for example, the span is 24 feet and the deck is 10 feet. Ten subtracted from 24 equals 14 and 14 divided by 2 is 7. Therefore, the run is **7 feet**. The rise is shown to be **10½** feet and the pitch of the rafters is the rise (10½ feet) divided by twice the run (14 feet) which is ¾.

To find this on the steel square, take 7 inches on the tongue and 10½ inches on the blade and make marks *A* and *B*, Fig. 29, at these points on a large piece of brown paper after having made a line along the outside edge of the tongue. Now lift the square and draw a line joining the two marks *A* and *B*. Now apply the square again to the paper but instead of placing the 7-inch division of the tongue at the point *A*, Fig. 29, place the 12-inch division of the tongue at

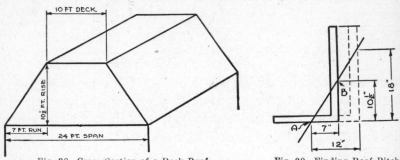

Fig. 28. Cross Section of a Deck Roof        Fig. 29. Finding Roof Pitch

this point, keeping the outside edge of the tongue on the same line as before, so that the square will now be in the position shown by the dotted lines in Fig. 29. When this is done, it will be found that the outside edge of the blade will cross the line *AB* (extended) at the 18-inch division. This shows that a pitch of 10½ inches in 7 inches, is the same as a pitch of 18 inches in 12 inches. Now a glance at the diagram shown in Fig. 25 shows that the pitch corresponding to a rise of 18 inches in 12 inches run or per foot run is a ¾ pitch. The same result could have been obtained by mathematics, but this exercise shows how the steel square may be used in many cases to avoid mathematics. In problems involving fractions of an inch, it may be quicker and easier to use the steel square.

**Types of Rafters.** As one of the principal uses of the square is in the laying out and cutting of roof rafters, it may be well worth while to study the names by which different kinds of rafters are known generally in the trade. For this purpose, refer to Fig. 23, which shows nearly all types of rafters used in roof framing.

The plate is the horizontal member extending all around the building at the top of the side and end walls at the place where the sloping rafters rest on these walls. The ridge or the ridge board is the horizontal member at the very top or peak of the sloping roofs against which the top ends of the sloping rafters rest. The common rafters rest on the plate at the bottom and slope upward from the plate to the ridge.

Hip rafters are the heavier members which slope diagonally up from the outside corners of a hip-roofed building to meet at the ridge level some distance back from the end wall. The hip rafters are made heavier than the common rafters because they are called upon to support the upper ends of a number of other rafters which are located at or near the ends of the building. Valley rafters are also heavier than common rafters, but they slope up from the plate to the ridge at the points where the wall plate of a projecting ell joins the wall plate of the main building. They are called valley rafters because rain water from the roof on each side flows down toward them. They are made heavy because they support the lower ends of a number of other rafters.

Rafters which do not extend all the way from the wall plate up to the ridge, but are supported either at the upper end by a hip rafter or at the lower end by a valley rafter, are called jack rafters. Rafters which have the lower end on the wall plate and the upper end against a hip rafter are called *hip jacks*. Rafters which have the lower end against a valley rafter and the upper end against the ridge are called *valley jacks*. In some roofs there are rafters which have the lower end resting against a valley rafter and the upper end resting against a hip rafter. These are called *cripple jacks* because they do not rest either on the wall plate or the ridge board. Jack rafters are of course shorter than the common rafters and hip and valley rafters are longer.

All the different kinds of rafters which go to make up such a roof frame as that shown in Fig. 23 must be cut to the proper length, and the ends must be cut and shaped correctly, so that they will all fit together to form the roof frame.

**Rafter Cuts.** The various cuts which must be made at the ends of the rafters to make them fit into their places have been given names by which they are known in the trade. These names follow.

The top or plumb cut is the cut which must be made in the end of a rafter where it rests against the ridge board. The bottom or heel cut is the cut which must be made in the end of a rafter where it rests on the wall plate. This is sometimes also called the foot cut or the seat cut. For common rafters these cuts are all made square across the rafter perpendicular to the sides of the rafter, although they are not at right angles to the length of the rafter. Hip and valley and jack rafters have to rest against surfaces which are not at right angles to them even in plan, and so these rafters, after hav-

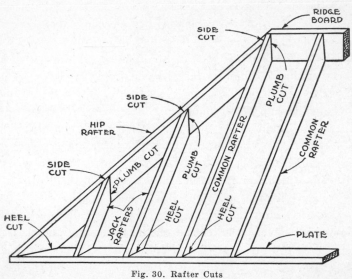

Fig. 30. Rafter Cuts

ing the top and bottom cuts made, must be further shaped at the ends so that they will fit against the other members which are diagonal to them in the plan of the roof. The ridge, for example, is diagonal to the hip rafters and valley rafters, while the hip and the valley rafters are diagonal to the jack rafters. These cuts are called *side cuts* and the sides of the hip and valley and jack rafters are said to be bevelled to fit against the other members. The side cuts are also sometimes called *cheek cuts*. Fig. 30 illustrates some of the different kinds of rafters and the cuts required on the ends of each one.

We have seen that the measuring line is a straight line which might be drawn on the side of a rafter parallel to the top edge of the rafter and reaching from the top outside edge of the wall plate to the center of the ridge. Fig. 24 shows a rafter with the lower end resting on the wall plate at *A*. The horizontal distance from the bottom edge of the rafter to the outside edge of the wall plate *P* (usually four inches) is the *seat* of the rafter on the wall plate and it is the length given to this seat which determines the position of the measuring line with relation to the top and bottom edges of the rafter. If the seat is long, the measuring line will be farther

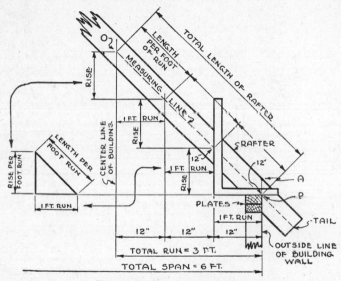

Fig. 31. Finding Length of Rafter

away from the bottom edge of the rafter and nearer to the top edge of the rafter than it would be if the seat were shorter. If a line is drawn on the side of the rafter representing the seat of the rafter on the top of the wall plate, then the point in which this seat line meets the measuring line is the point where the line of the top outside corner of the wall plate intersects the rafter (point *P* in Figs. 24 and 31).

What is called by the trade the length of the rafter is the distance taken along the measuring line from the outer edge of the wall plate (Point *P* in Figs. 24 and 31) to the center line of the ridge (point *O* in Fig. 31).

Figs. 23 and 30 show the rafters resting on the plate but not

overhanging it, that is, they show the frame of a roof with no projecting eaves and in this construction the measuring line is at the
top edge of the rafter, but both Figs. 24 and 31 show the rafters
projecting beyond the outside of the building wall so as to form projecting eaves. This projecting part of the rafter is called the *tail* or
*eave* and, although it is actually a part of the rafter, it is not figured
as a part of the length referred to in the preceding paragraph. The
tail of the rafter may be any length, to suit the ideas of the designer
of the roof, and the greater the projection of the eaves beyond the
outside of the building wall, the greater will be the overhang of the
eaves or tail.

**Laying out a Common Rafter.** Fig. 25 shows the steel square
with the blade upright and the tongue horizontal; the 12-inch run

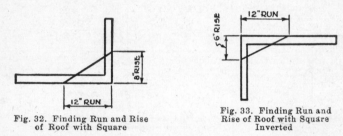

Fig. 32. Finding Run and Rise
of Roof with Square

Fig. 33. Finding Run and
Rise of Roof with Square
Inverted

is taken along the tongue, while the rise is taken along the upright
blade. It is just as correct and just as easy to take the 12-inch run
along the blade placed in a horizontal position, as shown in Fig. 32,
and the rise along the upright tongue. We may also reverse the
position of the square as shown in Fig. 33, still keeping the 12-inch
run on the blade with the blade horizontal and the rise on the tongue.
Now let us assume that we have to make a rafter in a roof of ⅓ pitch
which, as we have seen from Fig. 25, means that the rise of the rafter per foot of run is 8 inches. Let us assume that the rafter is 2
by 8 inches and that the span of the roof is 24 feet. We will also
assume that we have selected a stick of lumber more than long
enough to make the rafter. (How this may be done will be explained later.)

We now have our stick of lumber and our steel square. In
order to be able to make marks on the wood by which to cut it,
many carpenters use a scratch awl, such as is shown in Fig. 34, in-

stead of a pencil. This is a pointed steel rod about ¼ inch in diameter and 4 inches long, set into a wooden handle so that the whole tool is about 6 inches long. The sharp end is used like a pencil to make marks on the wood.

As rafters are usually laid out and cut at the site of the building, and as they are large and heavy and comparatively hard to handle,

Fig. 34. Scratch Awl
*Courtesy of The Stanley Rule and Level Plant, New Britain, Conn.*

it is customary to set up two horses as shown in Fig. 35, and to lay the stick of lumber which is to be cut into a rafter flat on its side on top of the two horses as shown, so that it is at a convenient height to work over and so that a carpenter when bending over his work can imagine himself to be looking at the side of the rafter. One edge of the stick of lumber previously should have been made straight and smooth by means of a plane and this edge will be the back or top edge of the rafter.

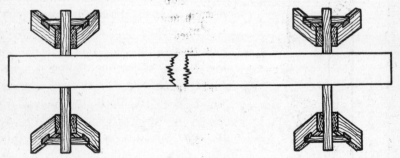

Fig. 35. Sawhorses

Many carpenters, in laying out rafters, work with the back of the rafter *toward* them, but it will be easier to understand the explanations if we work with the back *away* from us as the stick of lumber lies before us on the horses, and so we will place the stick of lumber on the horses with the straightened edge, which will be the top edge of the rafter, *away* from us. Let the right-hand end of the stick be the foot of the rafter at the eave of the roof. Assume an overhanging eave or tail of about 16 inches and, to allow for the

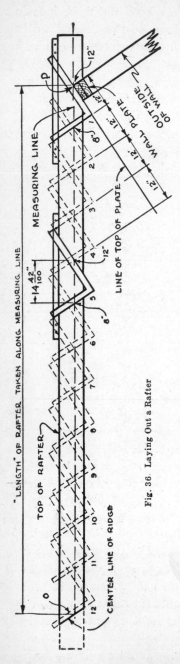

Fig. 36. Laying Out a Rafter

tail of the rafter, make a mark $P$, Fig. 36, about 17 inches from the right-hand end of the stick and about 5 inches away from the smoothed edge or top of the rafter. This mark or point represents the outside upper corner of the wall plate. With the aid of the steel square used as a rule, and a straight-edge, make a mark along the full length of the stick of lumber, passing through the point $P$ just made and keeping always the same distance (5 inches) away from the smoothed edge of the stick. This will be our measuring line.

Now, take the steel square and arrange the fence on it as explained in connection with braces. Lay the square on the stick with the 12-inch mark on the blade at the point $P$, and the 8-inch mark on the tongue exactly on the measuring line as shown by the full lines in Fig. 36. With these marks on the measuring line, hold the fence tight against the straightened edge of the stick and clamp it tightly in place by means of the screws. All this must be done carefully as an error at this point means that the subsequent work will be seriously wrong. With the fence held tightly against the dressed edge of the lumber and the 12-inch mark on the blade and the 8-inch mark on the tongue still exactly on the measuring line, take the scratch awl and clearly mark the point where the outside edge of the tongue of the square crosses the measuring line.

Keeping this point clearly in mind and with the fence held tightly against the dressed edge of the lumber, that is, against the top of the rafter, shove the square toward the left until the 12-inch mark on the outside edge of the blade is at this point and the 8-inch mark on the outside edge of the tongue is also on the measuring line but further along toward the left than it was before. The square is now in the position marked *2* in Fig. 36 and has been moved 14⁴²/₁₀₀ inches along the measuring line, because 14⁴²/₁₀₀ inches is the exact distance in a straight line between the 12-inch mark on the outside edge of the blade and the 8-inch mark on the tongue of the steel square.

Repeat this process, placing the square in the position marked *3* in Fig. 36; then do it again, placing the square in the position marked *4*. You will note that every time the square is moved 14⁴²/₁₀₀ inches to the left along the measuring line, it is advanced 12 inches, or one foot, in the direction parallel to the level of the wall plate, or in other words, parallel to the run of the rafter. The square will have to be moved along to the left just as many times as there are feet in the run. In the case of this particular rafter, it will be 12 times, since the run is one-half of the span and the span is 24 feet.

After the square has been moved along twelve times, make a mark along the outside edge of the tongue. This mark will indicate the center line of the ridge against the side of which the top of the rafter is to rest. The distance from this mark to the point *P* (the outside upper corner of the wall plate), measured along the measuring line, is the length of the rafter as indicated in Fig. 31, and the 14⁴²/₁₀₀ inches referred to is the length per foot of run. The total length will be 12 times 14.42 inches, which is 173.04 inches, or 14.42 feet, which is 14 feet and 5 inches for a span of 24 feet, a run of 12 feet and a rise per foot run of 8 inches.

If the span had been 20 feet, the run would have been 10 feet and the total length would have been 10 times 14.42 inches, which is 12 feet and ⅛ inch. If the rise per foot run had been 6 inches instead of 8 inches, and the span 20 feet, the total length would have been 10 times 13.42 inches, which is 134.2 inches or 11.18 feet, which is 11 feet 2 inches.

By moving the square along the rafter, as shown in Fig. 36,

and laying it on the rafter one-half as many times as there are feet
in the width of the building, the total length of the rafter may be
found without any multiplying or dividing, but the method is not
very accurate, as a little carelessness or a small mistake is increased many times and becomes serious. If the rafters are not
cut to just the right length, the roof frame will not fit tightly together and the structure will not be firmly braced. Therefore, the
method using the length per foot run is to be preferred. The rafter
tables of the steel square are of great help in using this method.

**Use of the Rafter Tables.** The best steel squares have on the
face of the blade or body a rafter or framing table, shown in Fig. 37,

| 2|2 | | 2| | | 2|0 | | |9 | |8 | |7 | |6 | |5 | |4 | |3 | |2 | |1 |
|---|---|---|---|---|---|---|---|---|---|---|---|
| LENGTH  OF  MAIN  RAFTERS  PER  FOOT  RUN | | 21 63 | 20 81 | 20 00 | 19 21 | 18 44 | 17 69 | 16 97 | 16 28 |
| "   HIP OR  VALLEY   "   "   "   " | | 24 74 | 24 02 | 23 32 | 22 65 | 22 00 | 21 38 | 20 78 | 20 22 |
| DIFFERENCE  IN  LENGTH OF JACKS 16 INCHES CENTERS | 28 84 | 27 74 | 26 66 | 25 61 | 24 585 | 23 588 | 22 625 | 21 704 |
| "   "   "   "   " 2 FEET | | 43 27 | 41 62 | 40 00 | 38 42 | 36 88 | 35 38 | 33 94 | 32 56 |
| SIDE  CUT  OF JACKS | | 6-11/16 | 6-15/16 | 7-3/16 | 7-1/2 | 7-13/16 | 8-1/8 | 8-1/2 | 8-7/8 |
| "   "  HIP OR VALLEY | | 8-1/4 | 8-1/2 | 8-3/4 | 9-1/16 | 9-3/8 | 9-5/8 | 9-7/8 | 10-1/8 |

Fig. 37. Rafter Tables
*Courtesy of The Stanley Rule and Level Plant, New Britain, Conn.*

which gives directly the length per foot of run for a large variety
of common rafters from those having a rise of 2 inches per foot of
run up to those having a rise of 18 inches per foot of run, as well as
other tables relating to hip and valley and jack rafters, which will
be explained later.

To find the lengths of common rafters per foot of run, look on
the first line below the outside edge of the blade, see Fig. 37, which
is marked *length of main rafters per foot run,* and under each of
the numbers in the inch line on the top edge of the blade of the
square from 18 inches down to 2 inches will be found numbers giving, in inches and hundredths of an inch, the length per foot run
of rafters whose rise per foot run is 2 inches, or 18 inches, or any
number of inches in between. Then, to find the length of the common rafter, multiply this length per foot run by the number of feet
in the run, which is one-half of the distance in feet across the building from outside to outside of the wall plates.

*Example.* Find the length of a common rafter in a roof where
the rise per foot of run is 8 inches and the width of the building from
the outside edge of the wall plate on one side to the outside edge

of the wall plate on the other side is 20 feet. This roof will have a one-third pitch. The run will be one-half of 20 feet or 10 feet.

First, find the rafter tables on the face of the steel square blade, see Fig. 37. Locate on the inch line along the outside edge of the blade the figure which is the same as the rise per foot run of the roof. In this example, the figure will be 8, since the rise per foot run is 8 inches. See Fig. 38A. Next, look on the first line under the figure *8* and you will find *14.42*, which means that the length of the

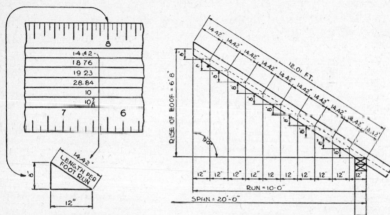

Fig. 38A. Use of Rafter Table          Fig. 38B. Length per Foot Run

rafter per foot run is 14.42 inches. Since the run is 10 feet and the length of the rafter per foot run is 14.42 inches, the "length" of the rafter will be 10 × 14.42 inches, which is 144.2 inches. By dividing by 12 this is found to be 12.01 feet, as shown in Fig. 38B; call it an even 12 feet.

It has been shown that by means of the rafter tables on the steel square, the length of a rafter can be found directly, without having to apply the square to the piece of lumber and move it along from point to point as was shown in Fig. 36. Having found the length of rafter from the rafter tables, it is possible to locate, on the side of the piece of lumber, the measuring line and the point where this measuring line will intersect the line of the outside of the wall plate (point *P* in Fig. 39) also the point where the measuring line will intersect the center line of ridge (point *O* in Fig. 39).

The measuring line is located far enough from the top edge or back of the rafter to leave the desired width for the tail, and at the same time the necessary length of seat for the rafter to rest properly on the wall plate. Having selected the location on the side of the rafter for the point $P$, the point $O$ (the intersection of the center line of ridge with the measuring line) can be located by measuring off along the measuring line the distance corresponding to the length of rafter (in this case, 12.01 feet) which was found by means of the rafter tables on the steel square.

Should the workman wish to use the top edge or back of the rafter as a measuring line, he can locate the point $A$, in which the line of the outside of the wall plate intersects the back of the rafter, by means of the steel square as shown at the right in Fig. 39. The

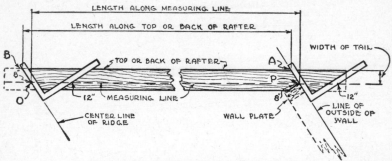

Fig. 39.  Length Measured along Back of Rafter

square is placed on the stock with the figure on the outside edge of the tongue corresponding to the *rise per foot run* of the roof (*8* in this case) on point $P$ and the square is moved around this point until the *12* on the outside edge of the blade rests on the measuring line, while the *8* on the tongue still rests at the point $P$. Then a line drawn along the outside edge of the tongue will be the line of the outside of the wall. The point $A$ will be the point in which this line intersects the line of the back of the rafter. Having located the point $A$, the length of the rafter (which has previously been obtained from the rafter tables) can be measured off along the back of the rafter and the point $B$ can thus be located. Point $B$ will be on the center line of the ridge, as is also point $O$. Of course the distance measured off for the length of the rafter along the measuring line and along the back of the rafter is the same in both cases and must be so because the *center line of the ridge* in Fig. 39 is necessarily parallel to the *line of the outside of wall*.

Fig. 40 illustrates how the *heel cut* or *seat cut* is made and also how the *top cut* or *plumb cut* is made. At the left is shown the heel cut. After point *P* is located on the measuring line as previously explained, to mark the point of intersection of the line of the outside of the wall plate, the square is placed on the stock as shown,

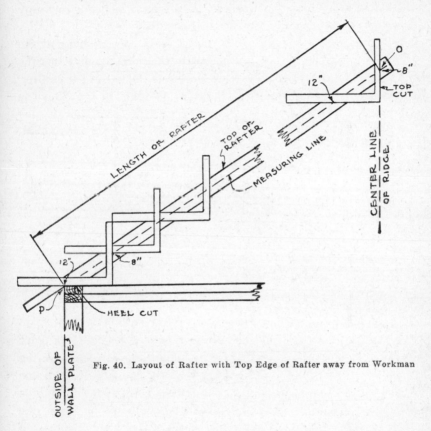

Fig. 40. Layout of Rafter with Top Edge of Rafter away from Workman

with 12 on the outside edge of the blade at point *P* and 8 (which is the rise per foot of run for the roof in this case) on the outside edge of the tongue of the square resting on the measuring line. Then a line drawn along the outside edge of the blade from point *P* to the lower edge of the rafter gives the heel cut. For the top cut the square should be moved along to the other end of the rafter, so that the 8-inch mark on the outside edge of the tongue will rest at the point *O* and the 12-inch mark on the outside edge of the blade will rest on the measuring line. Then a line drawn along the outside edge of the tongue will give the top cut or plumb cut and will be the line of the center line of the ridge. Allowance must be made for the thickness of ridge board, as is explained later.

Fig. 40, shows work being done with the back or top edge of the rafter away from the worker. As we have said before, many

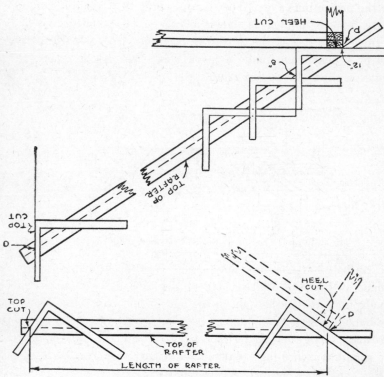

Fig. 41. Layout of Rafter with Top Edge of Rafter toward the Workman

carpenters prefer to work with the back of the rafter *toward* them. If the work illustrated in Fig. 40 were done in this way, it would look as shown in Fig. 41, which, as will be seen, is simply Fig. 40 turned upside down. Turn the page upside down and you will see that Fig. 41 is the same as Fig. 40.

Some carpenters also prefer not to use a measuring line, but to use the dressed edge or back of the rafter instead. Fig. 42 shows how this can be done for a rafter in a roof of ⅜ pitch with a span

of 26 feet and with a rise per foot run of 9 inches using the square
alone without the aid of a fence.  Refer back to Fig. 39.  Turn page
upside down and compare with Fig. 42.

This figure also shows what to do when the total run of the
rafter is not an even number of feet.  Suppose that the span is 27
feet, then the run is one-half of 27 feet, which is 13 feet and 6 inches.
Points $P$ and $A$ should be located as explained for Fig. 39 allowing
for the projecting eaves.  The square should be applied to the back
of the stick of lumber as shown in Fig. 42 at $A$ and moved along
12 times until it is in the thirteenth position.  Now take the scratch
awl or a pencil and make a mark along the outside edge of the
tongue as shown in the Fig. 42 marked "top cut for 13 ft."  Then with

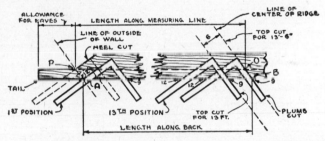

Fig. 42.  Heel Cut and Plumb Cut Using Top Edge as a Measuring Line

the 12-inch mark of the blade or body and the 9-inch mark of the
tongue still on the dressed edge of the stick of lumber, move the
square forward (to the right in the figure) until the 6-inch mark on
the outside edge of the blade or body comes directly over the mark
just made along the edge of the tongue.  This is the same as moving
the square 6 inches farther along *in the direction of the run of the
rafter,* making this run 13'6" instead of 13 feet.

In the same way, if the span were 20 feet and 8 inches out-to-out
of wall plates, then the run would be 10 feet and 4 inches and after
applying the square to the stick of lumber and moving it along 9
times to the tenth position, it should be moved along again until
the 4-inch mark on the outside of the blade or body is on the line
previously drawn along the outside edge of the tongue when it was
in the tenth position.  The same method can be followed to get the
length of a rafter with any other number of feet and odd inches in
the run of the rafter.

Fig. 43 shows the plan of a roof having a span of 27 feet in the main part, with an ell having a span of 20 feet 8 inches.

In Fig. 42 if the square in its first position at the left-hand end of the rafter were to be moved to the left (still keeping the 12-inch mark of the blade or body and the 9-inch mark of the tongue, on the dressed edge of the lumber) until the outside edge of the body passes through the point $P$ which represents the upper outside corner of the wall plate, then a line drawn along the outside edge of the blade or

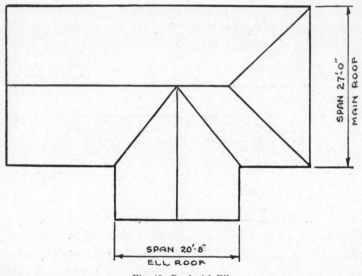

Fig. 43. Roof with Ell

body of the square and passing through the point $P$ would give the *heel cut* or *seat cut* at the lower end of the rafter where it rests on the wall plate.

Allowance was made in the length of the lumber as shown in Fig. 42 for the tail of the rafter which projects out beyond the outside edge of the wall plate to form the overhanging eaves.

In a similar way, if, in Fig. 42, when the steel square is in its final position at the right-hand end of the stick of lumber (the top of the rafter) a line were drawn on the side of the lumber along the outside edge of the tongue marked *plumb cut* in Fig. 42, this line would represent the center line of the ridge. If there were no ridge board and the rafters on each side of the roof simply butted against each other, this line would be the line on which to cut the lumber at this (the top) end. If there is a ridge board, the steel square must be moved *back* (to the left in Fig. 42) a distance equal to one-half of the thickness of the ridge board as follows: with the 12-inch mark on the outside edge of the blade or body of the square and the 9-inch mark on the tongue of square both still on the dressed edge of the lumber, move the square to the *left* (away from the ridge line) until the distance marked *6* in Fig. 42 is reduced by one-half the thickness of the ridge board, and then a line drawn along the outside edge of the tongue of the square will be the line on which to cut the lumber for the upper end of the rafter. As this may not be quite clear in Fig. 42 where the top edge of the rafter is *towards* the worker, Fig. 44 is given, showing the top edge of the rafter *away*

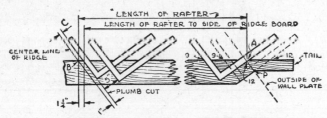

Fig. 44. Making Allowance for Half of Ridge

from the reader and showing the steel square in the two positions, at the right for the seat cut and at the left for the plumb cut. Figs. 44 and 42 are essentially the same except that in Fig. 44 allowance is made for half the thickness of the ridge board.

In this figure the thickness of the ridge is assumed to be 2 inches and when the square is moved back so that the distance $C$ perpendicular to the top cut mark or plumb line is one-half of this, or 1 inch, it will be found that the square has been moved $1\frac{1}{4}$ inches along the measuring line or along the top edge of the rafter. This makes the length of the rafter after deduction for ridge $1\frac{1}{4}$ inches less. Thus we see that one-half the thickness of the ridge board is not measured along the measuring line, but in a direction at right angles to the plumb cut or top cut.

Fig. 44 is the same as Fig. 42 turned upside down, except that it shows the square moved back to allow for the ridge board.

In order to make more clear the reason why it is necessary to allow for one-half the thickness of the ridge board, Fig. 45 is given showing a large scale section through the ridge. Compare this figure with the left-hand end of Fig. 44.

Since the common rafters in a roof are always at right angles to the ridge and the wall plate, both the top cut and the seat cut are made at right angles to the sides of the rafter. This makes the laying out and cutting for common rafters much more simple than for hip and valley and jack rafters, the top cuts for which have

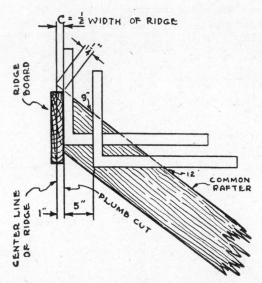

Fig. 45. Section through Ridge Showing Common Rafter

to have a *side cut* in addition to the *plumb cut* to make them fit against each other, as will be seen by reference to Fig. 30.

**Hip and Valley Rafters.** It has already been pointed out that the hip rafters are the heavy rafters which slope up and back from the outside corners of a hip-roofed building to the ridge and that the valley rafters are similar heavy rafters which also slope up from the outside wall to the ridge, but which start from the angle between the main roof and a projecting ell. If you were to look straight down onto the top of a roof, as from an aeroplane, you would see the roof in plan, and if the building were still under construction so that the rafters were all in place but not as yet covered up by the roof boarding, you would be able to see all of the rafters "in plan" as shown in Fig. 46, where *cd* and *ce* are hip rafters, and *gh* is a valley rafter.

If you were to look at this same roof from a position lower down and a little to one side, as for instance, from the top of a high tree

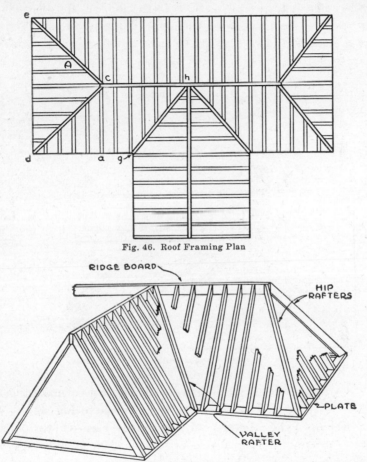

Fig. 46. Roof Framing Plan

Fig. 47. Hip and Valley Rafters

near by, the framing might look something like Fig. 47. It will be seen that the total rise of the hip and valley rafters is the same as the total rise of one of the common rafters.

In the simplest roof of this type, the slope of the hipped portion *A* in Fig. 46 is the same as the slope of the side of the main roof, and the perpendicular distance of the point *c* from the end *de* and from the side $dg$ is the same. This is called a roof of equal pitch. In such a roof frame the hip rafter in plan, see Fig. 46, forms the diagonal of a square of which the common rafters meeting at the ridge point $C$ form (in plan) two of the sides, while the side and end

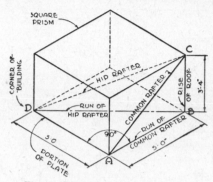

Fig. 48. Diagram of Relative Positions

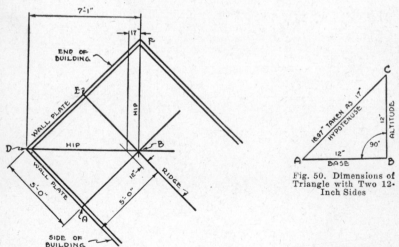

Fig. 49. Plan of Roof Showing Run of Hip Rafter

Fig. 50. Dimensions of Triangle with Two 12-Inch Sides

walls of the building form the other two sides. Now since the rafters do not lie flat as they appear to do in a plan view, but actually all slope upward as seen in Fig. 47 and in the diagram of relative positions, Fig. 48, it would be more correct to say that the *run* of the hip rafter forms the diagonal of a square of which the *runs* of the common rafters form two of the sides while the wall plates at the side and end of the building form the other two sides. Fig. 49 shows this condition in plan turned so that you are looking directly down on the run of the hip rafter *DB*. Now remember that point *B* is an imaginary point at the level of the top of the wall plate and that the *runs* of the common rafters *AB* or *EB* and the *run* of the hip rafter *DB* meet at this point. Thus the run of the hip rafter *DB* forms the

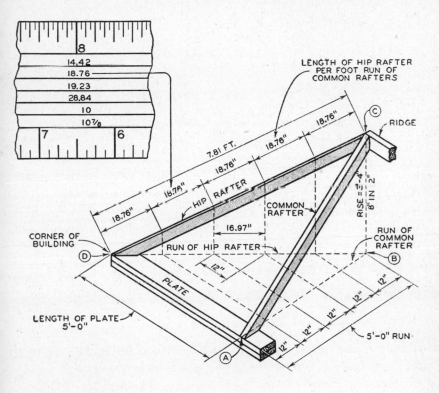

Fig. 51. Relative Dimensions of Hip and Common Rafters

diagonal of a square *A B E D*, and *DB* is the hypotenuse of the right-angled triangle *DAB*. In every right triangle of this kind where the two sides are equal, it is a fact that for every 12 inches of run along the sides there is a run of 17 inches along the hypotenuse, see Fig. 50. This means that in every hipped roof of equal pitch (such as is shown in Fig. 51), for every foot of *run* of the common rafter, the hip rafter has a *run* of 17 inches. This is a very useful fact to remember.

It has been seen that the point $B$ is not on any of the rafters nor is it on the ridge, because the rafters all slope upward from the wall plates to the ridge which is at a height above point $B$ exactly equal to the total rise of the roof or of the rafters. Also, although the run of the hip rafter is $17/12$ times the run of the common rafter, the total rise of the hip rafter is exactly the same as the total rise of the common rafter. The rise of the hip rafter at any point is the same as the rise of the common rafters at the corresponding point. Thus, if the pitch of the common rafter is 9 inches per foot run, the pitch of the hip rafter is 9 inches per 17-inch run; if the pitch of the common rafter is 6 inches per foot run, the pitch of the hip rafter is 6 inches per 17-inch run; if the pitch of the common rafter is 8 inches per foot run, the pitch of the hip rafter is 8 inches per 17-inch run, and so on. This is illustrated in Fig. 51.

A tradesman usually starts to lay out a roof frame knowing the span of the roof out-to-out of wall plates and the total rise of the roof. From these he gets the rise per foot run of the common rafters by the methods already explained. It is then necessary to find the relation between the length of the hip rafter and the rise per foot run of the common rafter.

The first thing to do is to find the length of the hip rafter per foot run of the common rafter and then, by multiplying this length by the run of the common rafter (half the span of the roof), the length of the hip rafter is found, provided that the roof is a roof of equal pitch. To find this length of the hip rafter per foot run of the common rafter, take the rise per foot run of the common rafter on the tongue of the square and 17 on the blade of the square and with a rule, or another square, measure the distance between these two points, as shown in Fig. 52. For a rise per foot run of common rafter of 8 inches, this distance will be 18.76 or 18¾ inches and this, multiplied by one half of the roof span will be the length of the hip rafter from the outside upper corner of the wall plate to the center line of the ridge. Fig. 51 gives an illustration of this fact and it also shows a way to get the length of hip rafter per foot run of common rafter directly from the steel square.

Look on the face of the blade of a steel square which has the rafter tables, Fig. 37, and in the *second* line indicated as "Length of Hip or Valley Rafter per foot run" will be found figures indicating in inches and hundredths of an inch the length of hip rafters per foot run of common rafter for roofs whose rise per foot run of the common rafter is indicated by the figure on the inch line at the head of the column of figures.

*Example.* To find the length of a hip rafter for a roof of equal pitch where the rise per foot run of the roof is 8 inches and the span of the building is 10 feet: first, find on the inch line on the outside edge of the blade of the steel square the figure *8*, which represents the 8-inch given rise per foot. Now, look in the column of figures marked on the side of the square underneath the figure *8*, as shown in Fig. 51, and in the *second line* underneath the *8*, will be seen the

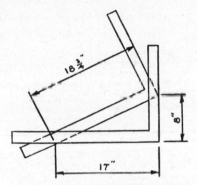

Fig. 52. Length of Hip Rafter per Foot Run of Common Rafter

figure *18.76*. This is the length of the hip rafter in inches (eighteen and seventy-six one hundredths), for every foot in the run of the common rafter, or, to put it in another way, for every foot in the half-span of the roof out-to-out of wall plates. Since the whole span of the roof is 10 feet, the half-span or the run of the common rafter is 5 feet, and the *length* of the hip rafter is 5 × 18.76 inches, which is 93.8 inches. Dividing this by 12 gives 7.81 feet, which is very near to 7 feet 9¾ inches. This example is illustrated in Fig. 51.

If the rise of the common rafter per foot of run were 9 inches instead of 8 inches, that is, if it were a ⅜ pitch roof instead of ⅓ pitch, you would proceed in the same way, but would look at the figure in the *second* line under the figure 9 on the face of the blade of the square, where you would find the figure 19.25. This is the length of the hip rafter in inches per foot run of the roof. If the span of the roof were 27 feet, the length of the hip rafters would be 13½ × 19.25 inches, which is very nearly 21 feet 7⅞ inches. This length may be used directly to avoid stepping the square along as explained in the following.

To lay out the hip rafter, proceed just as in laying out the common rafter in Fig. 42, except that where, in the case of the common rafter, you took 12 inches on the blade and 9 inches on the tongue of the steel square, in the case of the hip rafter take 17 inches on the blade and 9 inches on the tongue as shown in Fig. 53. Having established the length of the hip rafter by means of the steel square as explained, choose a stick of lumber a little longer than this length to provide for the tail of the rafter at the eaves and place the

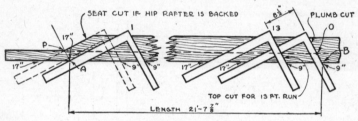

Fig. 53.  Layout for a Hip Rafter

square on the piece of lumber with **17** on the blade and **9** on the tongue, coinciding with the dressed top edge of the rafter which is *towards* us in this figure.  The measuring line *PO* can be drawn parallel to the dressed edge or back of the rafter and far enough from this edge so that the distances *PA* or *OB* *measured along the outside edge of the tongue of the square* will be the same as the distances *PA* or *OB* for the *common rafter* shown in Fig. 42, also measured along the outside edge of the tongue of the square.  (Note that the point *P* is, in both cases, a point on the line of the outside upper edge of the wall plate; the line *PA* is in both cases a line in the plane of the outside face of the wall plate, and the point *A* is in both cases a point on the line of intersection between the plane of the outside face of the wall plate and the plane of the sloping surface of the roof, where the roof has overhanging eaves as illustrated in Fig. 31. Where the roof has no overhanging eaves, as shown in Fig. 51, the measuring line coincides with the top edge or back of the rafters and the points *P* and *A* are the same.)

The measuring line is used in this case only for the purpose of locating the point $P$, but the actual measuring is done along the back of the rafter. Of course the measuring could be done along the measuring line if desired. In actual work the entire measuring line would not be drawn out on the lumber, but a short piece of it would be drawn parallel to the dressed edge of the rafter at the *approximate* location of the point $P$.

Of course the entire measuring line could be drawn out parallel to the dressed edge and the point $O$ located on it by measuring off the length 21 feet $7\frac{7}{8}$ inches along the measuring line. To avoid drawing out the entire measuring line, the dressed edge or back of the rafter may be used by proceeding as follows: The point $P$ can be located on the measuring line far enough from the left-hand end of the piece to allow for the tail, and the square is put in the position for the seat cut as shown. Draw a line from $P$ at right angles to the seat cut, locating point $A$ on the back or top edge of the rafter and move the square to this new position. Point B can now be located by measuring off the length $21'-7\frac{7}{8}''$; or the square may be stepped along the rafter edge to the right 12 times as described for the common rafter in Fig. 42. This gives the position of the "top cut for 13-foot run" since the square has been applied to the top edge of the rafter 13 times. However, the run of the common rafter in this case is 13 feet and 6 inches, so find the position of the "top cut" for an additional run of 6 inches on the common rafter, remembering that this is not a common rafter, but a hip rafter. Fig. 51 shows that for every 12 inches in the run of the common rafter, there is a run of 17 inches in the hip rafter; therefore, for an additional run of 6 inches in the common rafter, there must be an additional run of $8\frac{1}{2}$ inches in the hip rafter. This is illustrated also in Fig. 54, which shows a square 12 inches on each side, representing 12 inches of run of a common rafter, and shows a diagonal $m$, 17 inches long, representing the corresponding run of the hip rafter. For every inch on the run $ab$, of the common rafter, there will be a run of $\frac{17}{12}$ inch in the hip rafter, so that for a run of 6 inches in the common rafter there must be a run of $\frac{17}{12} \times 6 = 8\frac{1}{2}$ inches in the hip rafter.

Referring back to Fig. 53, at the right the square is shown in position 13, giving the top cut for a 13-foot run of roof along the outside edge of the tongue. To get the top-cut line for a run of 13

feet and 6 inches, move the square along to the right (keeping the
17 on the blade and the 9 on the tongue always on the dressed edge
of the lumber) until the 8½-inch mark on the outside edge of the
blade is on the 13-foot top-cut line.   Then a line marked along the
outside edge of the tongue of the square in its new position will give
the angle for the top cut for a 13-foot 6-inch run and the approximate location of this cut.   To get the actual location of the cut, allowance must be made for ½ the thickness of the ridge board and for
the side cut on the rafter.   Fig. 55 shows the hip rafter and the
ridge board in place and the side cut at the ridge.

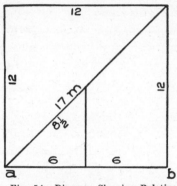

Fig. 54.  Diagram Showing Relative
Lengths of Runs for Hip and Common
Rafters in Equal-pitched Roof

Figs. 53 and 56 are really views of the hip rafter taken along
its center line just as though the rafter had no thickness or was
like a sheet of paper.   Therefore, the position of the square as
shown for the top cut is not correct for two reasons.   First, it is
shown passing through the line of intersection of the center of the
hip rafter, with the center of the ridge board and would have to be
moved back *to the left,* a distance equal to 1½ times half the thickness of the ridge board to allow for the thickness of this member.
Second, it would then not be correctly placed for the top cut because no allowance would be made for the thickness of the hip
rafter itself and for the *side cut* shown at the right-hand side of Fig.
55.   To allow for this, the square would have to be moved again,
this time *to the right,* a distance equal to half the thickness of the
hip rafter measured along the blade of the square.   If the hip rafter

is twice as thick as the ridge board, as is quite usual, this will bring the square back practically to the position shown in Fig. 56, as shown in the lower right-hand corner of Fig. 55.

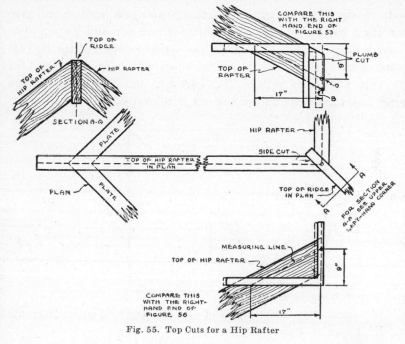

Fig. 55.  Top Cuts for a Hip Rafter

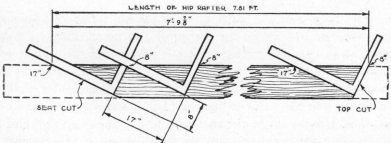

Fig. 56.  Laying Out a Hip Rafter without a Tail

Fig. 56 shows another method for laying out the hip rafter, this time assuming that the rafters do not overhang the wall plate (see Fig. 51), and assuming a run of common rafter equal to 5 feet, a rise per foot run of 8 inches, and working on the stick of lumber with the top edge of the rafter *away* from the workman.  First, place the square in position for the seat cut at the left-hand end of the piece of lumber, using the dressed top edge of the stock as the measuring line and with the 17 on the blade or body of the square and the 8 on the tongue of the square on this line. Then make the seat cut, there being in this case no allowance for a tail. Next, having determined from the rafter tables on the square that the length of the hip is 5 × 18.76 inches or 7.81 feet, or 7 feet 9⅞ inches, measure off this distance along the edge of the stock and make a mark to indicate that here is the center of the ridge. Now place the square again on the stick with the 17-inch mark on the body and the 8-inch mark on the tongue, both exactly on the top edge of the stock (which in this case is the top edge of the rafter) and with the 8-inch mark on the tongue far enough from the center of the ridge mark to allow for 1½ times one half the thickness of the ridge board, and for the thickness of the hip rafter and the side cut of this rafter against the ridge. Then a mark along the outside edge of the tongue of the square will be the top cut for the hip rafter. The allowance to be made for the ridge thickness and side cut in Fig. 53 will be about equal to the thickness of the hip rafter as shown in the upper right-hand corner of Fig. 55.

From a study of Fig. 47, it will be seen that the valley rafters are practically the same as the hip rafters and would be laid out and cut by the same methods.

The fact that both the hip and the valley rafters do not meet the ridge squarely but meet it at an angle, usually of 45° in plan means that they cannot be cut at the top at right angles to the sides of the rafters, but, in order to make them fit against the ridge properly, another cut, called the side cut, must be made after the top cut is located. This is illustrated in Figs. 30 and 55.

Without the aid of the rafter tables on the steel square, it is very puzzling to figure out just how this side cut should be made in order to give the hip and valley rafters a snug fit against the ridge. There have been a number of rules in use among tradesmen for making this cut, but the simplest method is to make use of the rafter tables and then to make the cut as shown in Figs. 57 and 58. Fig. 57 shows the hip rafter as it would set in the roof frame, *DE* being the top cut, or plumb cut, where the hip rafter meets the ridge board, and *CD* being the side cut, sometimes called the cheek cut.

Fig. 58 shows the piece of lumber from which the rafter is to be cut turned over on edge so that you are looking straight down on the top or *back* of the rafter. A line is drawn straight through the middle lengthwise from end to end to serve as a measuring line. The point in which the top cut in Fig. 56 intersects the top edge of the rafter is really on this measuring line in the center of the top edge, rather than on the side of the rafter. Fig. 55 shows this point transferred to the side of the rafter from the measuring line. At the upper end of the rafter (the right-hand end in Fig. 58), locate a point representing the top of the top cut. To make the side cut, apply the

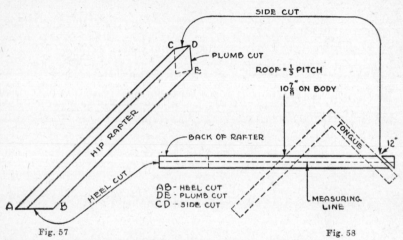

Fig. 57                                         Fig. 58

Figs. 57 and 58. Heel, Plumb, and Side Cuts for a Hip Rafter

steel square to the top edge of the rafter as shown in Fig. 58 with the 12-inch mark on the tongue exactly on the above-mentioned point. Keeping the 12-inch mark on this point, the square is to be moved about to take a position depending upon the slope or pitch of the roof in inches per foot of run.

Assume that the roof has a rise of 8 inches per foot of run. Look on the rafter tables on the face of the blade or body of the square and, in the column of figures underneath the 8-inch mark on the inch line at the outside edge of the blade, the sixth figure will be found to be $10\frac{7}{8}$ as shown in Fig. 38. Then find the $10\frac{7}{8}$-inch mark on the inch line at the outside edge of the blade or body of the square and place this mark directly over the edge of rafter as shown in Fig.

58, while the 12-inch mark on the outside edge of the tongue is still held at the point representing the upper end of the top cut. With the square in this position, a mark made along the outside edge of the tongue across the top edge of the rafter is the line of the side or cheek cut.

On the sixth line in the rafter tables on the face of the body of the steel square, will be found figures which are to be used for getting the side cuts of hip or valley rafters (just as the 10⅞ was used in the example explained) for roofs whose rise per foot of run corresponds to the inch marking at the head of the column. For example, 10⅞ occurs in the sixth line in the column underneath the 8-inch marking and was used in Fig. 58 for a roof with an 8-inch rise per foot of run.

The seat cut, or heel cut, for the hip rafter is made with the square held in the position shown in Fig. 56 if working with the back of the rafter *away* from the workman, or as shown in Fig. 53 if working with the back of the rafter *towards* the workman, but if, in Fig. 53, the point *P* represents the outside upper corner of the wall plate, then in making the heel cut, the square must be moved a short distance to the right to allow for the top edges of the rafter which would project above the line of the top of the common and jack rafters as shown in Fig. 63 unless the corners of the hip rafter are cut off or *backed* as it is known to the trade. The backing of the hip rafter is troublesome and so, in order to avoid doing it, the rafter is usually lowered by moving the square when making the heel cut. Moving the square in this way when the heel cut is made results in the hip rafter being slightly lowered so that the corners will not project above the general plane of the adjacent roof surfaces.

The distance which the square must be moved back depends upon the thickness of the rafter as illustrated in Fig. 59. This figure shows the wall plates at a corner of a building in plan and the top of the hip rafter looking straight down upon it where it rests on the wall plates. The thickness of the rafter is the distance *3–3*. Points *1* and *2* are on a line lengthwise along the center of the top of the rafter (the measuring line in Fig. 58). Point *2* is directly over the corner in which the outside edges of the wall plates in the two walls intersect. Points *3* are directly over the points in which the sides of the hip rafter intersect the outside edges of the two wall

plates.  If the top surface of the hip rafter is square with the sides
of the hip rafter, point *1* must be at the same height as points *3*.  But,
since the rafter slopes upward from point *2* towards the ridge, point
*1* must be higher up than point *2*.  Therefore, points *3* on the top

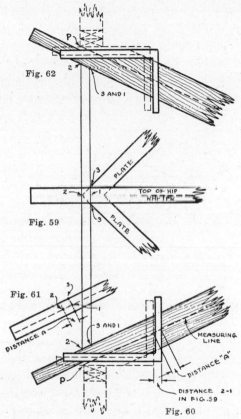

Figs. 59 to 62.  Making the Heel Cut On a Hip Rafter
to Avoid Backing

edges of the hip rafter must also be higher up than point *2*.  Then if
point *2* is in the roof surface, points *3* will project above the roof
surface because all points *in* the roof surface on lines parallel to the
outside edges of the wall plates are at the same elevation.  Points *3*
are on the top edges of the hip rafter and these edges cannot be
allowed to project above the roof surface, therefore the entire hip
rafter must be dropped down or lowered until points *3* are in the roof

surface. When this is done, point *2* will be slightly below the roof surface.

In order to lower the hip rafter, it is necessary to raise the level of the seat cut with reference to the measuring line on the side of the rafter, and to do this the steel square must first be placed in position with 17 on the blade on the measuring line at the point representing the outside upper corner of the wall plates (point *P* in Fig. 60) and instead of making the seat cut along the outside edge of the blade of the square in this position, the square should be moved to the right along the measuring line until the horizontal projection of the distance *2–1* (which, in the plan view, Fig. 59, is the same as half the thickness of the hip rafter) is measured off in the direc-

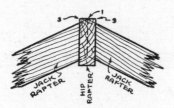

Fig. 63. Section through Hip Rafter Showing How It Projects above Tops of Other Rafters if It Is Not Dropped

tion of the seat cut or to the right of the point *P* in Fig. 60. In other words, the square should be moved *along the measuring line* towards the ridge a distance equal to distance *2–1* in Figs. 60 and 61, *measured along the top of the rafter*. With the square in this new position, the seat or heel cut can be made with the assurance that the edges of the hip will not project beyond the roof surfaces.

Fig. 60 shows the square in its first position by the dotted lines and in its second position by the full lines. This figure shows a side view of the hip rafter with the top *away* from the worker. The distance *A*, Fig. 60, by which the square is moved along on the measuring line varies, depending upon the slope of the roof and the thickness of the hip rafter. This distance will be somewhat more than the half thickness of the hip rafter. Actually it will be this half thickness divided by 17 and multiplied by the length of the hip rafter per foot run of the roof. It would be safe to move the square along ⅝ of the thickness of the hip rafter for ⅓-pitch roof, ⅔ of the thickness for ½-pitch roof, and ¾ of the thickness for steeper roofs. The square is *moved* along the measuring line and thus away from the seat cut, but it is moved *to the right* or *in the direction of the seat cut* a distance equal to distance 2-1 in Fig. 59. This results in bringing the outside edge of the body of the square (and consequently the new seat cut) nearer to the top edge of the rafter and when the new seat cut is made the result is that the whole rafter is lowered such a distance that points 3, Fig. 63, will lie in the general roof surface instead of projecting above it.

Fig. 61 shows a view of the top surface of the hip rafter.

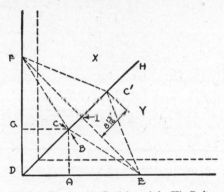

Fig. 64. Illustrating Position of the Hip Rafter

Fig. 62 shows a side view of the hip rafter with the top *towards* the worker.

**Backing.** Fig. 63 shows a section or end view of the hip rafter and illustrates another way of getting over the difficulty and preventing the top edges of the hip rafter from projecting above the roof surfaces. This is to cut these corners off so that the top edge of the rafter will be of such a shape that all parts of it will be in the roof surfaces on either side of the hip line and so that the roof boards will fit properly over the hip rafter.

Fig. 64 is a plan of part of a roof with a hip *DH* in which the sloping roof surfaces *X* and *Y* come together. Assume that the pitch of both of the roof surfaces is 8 inches to the foot, or, in other words, that it is a roof of equal pitch. Let the 2 lines *FGD* and *DAE* represent the outside faces of the 2 wall plates which come together at the corner, and let the dotted lines just inside of these represent the inside faces of the wall plates. Suppose that the distance *A B* in plan is just 12 inches; that is, that *A B* represents a 1-foot run of the common rafter and the point *C* is on the sloping hip line just 8 inches above the point *B* in plan, since the pitch of the roof surfaces is 8 inches to the foot. Thus point *C* is also 8 inches above points *A, D* and *G,* since these points are on the level outside top corner of the wall plates.

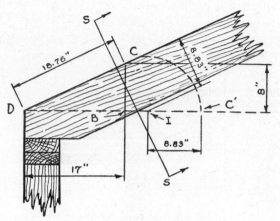

Fig. 65. Section through Roof along Hip Line

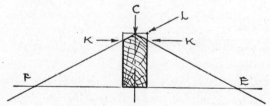

Fig. 66. Section through Hip Rafter at Right Angles to Hip Line

Now suppose that the line *FE* is drawn through the point *I* in plan at right angles to the hip line *DBH* and let line *FE* represent a plane or cut which is passed through the roof surfaces at right angles to the actual sloping hip line so that the plane or cut intersects the horizontal plane at the level of the top of the wall plates in the line *F I E*. If this plane or cut surface is revolved or rotated using the base line *F I E* as a fixed axis until the plane lies flat on the plan of Fig. 64, then the point *C* on the hip line will move to the point C', the distance IC' being 8.83 inches or $8\frac{13}{16}$ inches, as illus-

trated in Fig. 65 which represents a section through the roofs parallel
to the hip line $DC$ and looking directly at the side of the hip.  The
dimension 8.83 inches, for $CI$, is arrived at by comparing the dimensions of the various triangles shown in Fig. 65, which shows
that $CI$ is $\dfrac{18.76}{17} \times 8$, which is 8.83.

Now in the angle $FCE$, Fig. 66, is shown a small section through
the hip rafter which is sliced out of it by the plane or cut $FCE$ in
Fig. 64 and as is indicated by section $S$—$S$, in Fig. 65. Fig. 66 shows a
larger view of this section $S$—$S$, illustrating the end of the hip rafter

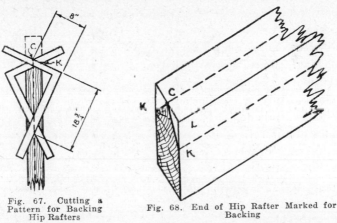

Fig. 67.  Cutting a
Pattern for Backing
Hip Rafters

Fig. 68.  End of Hip Rafter Marked for
Backing

or a section cut through it at right angles to its length.  It shows at
$CLK$ the corners to be cut off.

The size of these corners to be cut off for roofs of different
slopes can be found with the help of the steel square as follows: first,
take a dressed board the same width as the thickness of the hip
rafter, which may be 3 or 4 or more inches.  This board must have
squared ends and must be about 3 feet long.  The board is to serve
as a pattern from which to mark the end of the rafter, as the depth
of the rafter is not sufficient to allow for manipulation of the steel
square.

Having prepared the board which is to serve as a pattern, make
a line in the exact center of one side, running from end to end of the
board.  Now in the rafter tables on the face of the blade of the steel
square, find in the inch line at the outside edge of the blade, the
figure corresponding to the rise per foot run of the common rafter,

in this case 8 inches. In the column of numbers under this figure 8, look at the second number from the top which is *18.76* as shown in Fig. 51. Fig. 37 tells us that this number in the second line from the top is the length of the hip rafter per foot of run. Using the length of the hip per foot of run on the blade of the square and the rise per foot of run of the roof on the tongue of the square (8 inches in this case) place the square on the pattern as shown in Fig. 67 with these numbers on the center line as shown. Now mark along the outside edge of the tongue the line *C K* and, reversing the square as shown, mark in the same way on the other side of the center line. The pattern may be cut on these marks and applied to the end of the hip rafter to give the measurements for backing the rafter. Fig. 68 shows the end of a hip rafter with the lines marked on it ready for backing. When the rafter is cut along these lines, it will fit into the roof with no interference from projecting edges.

Theoretically, a similar problem arises in the case of valley rafters; that is, the top edge of the valley rafter, if it is square with the sides of the rafter, does not lie in either of the roof surfaces which intersect or meet in the valley. Theoretically, the top edge of a valley rafter should be gouged or hollowed out in the shape of a shallow **V**. Practically, however, it is found that it is not important to have the upper edge of the valley rafter exactly coinciding with the roof surfaces and, since the outside upper edges lie *below* the roof surfaces instead of above them as in the case of hip rafters, any necessary adjustment can be made more easily in the roof boarding than in the valley rafters themselves.

**Jack Rafters.** In every roof containing a hip or a valley there are some rafters similar to the common rafters, but shorter than these. They rest at top or bottom against the ridge or on the wall plate, but have the other end cut to rest against the hip or the valley rafter. These jack rafters are spaced at equal distances apart, the same as common rafters, usually either 16 inches or 24 inches center to center. Because they fill a triangle-shaped space in the roof surface, they vary in length. Starting with the shortest one at the top or bottom of the hip or valley, they increase regularly in length, the second jack rafter being twice as long as the first one, the third rafter three times as long as the first one, and so forth, as shown in Fig. 69. The jack rafters have the same rise per foot run as do the common rafters on the same slope of the roof, and the cuts where they rest on or against the wall plate or the ridge board are obtained in the same way as was explained for the common rafters. They differ from the common rafters as to length and in the side cut necessary to make them fit against the hip or valley rafter.

The lengths of jack rafters may be found by subtracting from the length of the common rafter an amount depending upon the distance of the center of the jack from the last full-length common rafter or from the end of the ridge. If the roof is a roof of equal

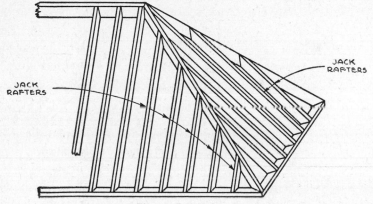

Fig. 69. Jack Rafters

pitch so that the hip rafter is at an angle of 45°, in plan, with the ridge and the wall plates (see Fig. 46), then the difference in length between the run of the jack and the run of the common rafter is 1 foot for each foot of distance between the center line of the jack rafter and the center of the last common rafter at the end of the ridge. This means that the difference in length between any jack rafter and the length of the common rafter in the same roof slope is equal to the length per foot run of the common rafter, multiplied by the distance in feet between the center of the jack rafter and the center of the last common rafter at the end of the ridge, because the two rafters are in the same roof surface.

To find the length of a rafter per foot of run for any pitch, place a rule diagonally across the steel square from the 12-inch mark on the blade of the square to the figure on the tongue representing the rise of the roof per foot of run. Measure with the rule the diagonal

distance between these two points, and this measurement will give
the length of the rafter that will span over 1 foot of run as shown
in Fig. 70 for a rise of 8 inches per foot of run. The measurement
is 14½ inches and this would be the difference in the length of
jack rafters if they were 1 foot apart center to center. If the jack
rafters were 24 inches apart, the difference in length for each one
subtracted from the length of the next longest one, would be 2 × 14½
inches, which is 29 inches.

Fig. 71 shows how to get the difference in length of jack rafters
by means of the steel square when they are 16 inches apart center

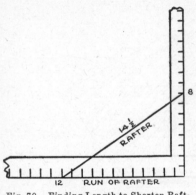

Fig. 70. Finding Length to Shorten Rafters for Jacks per Foot of Run

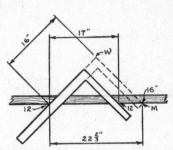

Fig. 71. Finding Difference in
Lengths of Jack Rafters Which Are
16 Inches on Centers with a 12-Inch
Rise per Foot Run

to center. Select the piece of lumber from which a jack rafter will
be cut, and dress one edge. Then lay the steel square on the piece
of lumber with 12 on the blade on the dressed edge and the rise
per foot of run for the common rafter on the tongue (in this case
12 inches) also on the dressed edge of the lumber. This will show
by measurement the length per foot of run to be 17 inches. Now, if
the spacing of the jack rafters is 16 inches center to center, move
the square, as shown, along the line of the blade until the blade
measures 16 inches and the tongue of the square would then be as
shown from $W$ to $M$, also 16 inches. The difference in the lengths
of the jack rafters would then be the measured distance from 16 on
the blade of the square to 16 on the tongue of the square measured
along the edge of the piece of lumber. It would be 22⅔ inches.

If the rise of the roof per foot run were 9 inches and if the
spacing of the jack rafters were 18 inches center to center, the same

method might be used for finding the amount by which each successive jack rafter is shortened. This is shown in Fig. 72.

As shown in the illustration, the steel square is placed on the piece of lumber from which the jack rafter is to be cut with the 12-inch mark on the outside edge of the blade and the 9-inch mark on the outside edge of the tongue resting on the dressed edge of the lumber which will be the top or back of the jack rafter. By moving the square forward along the line of the blade until the distance along the outside edge of the blade from the heel of the square, $W$, to the dressed edge of the lumber measures 18 inches, and the tongue of the square takes the position $WM$ in Fig. 72, the distance

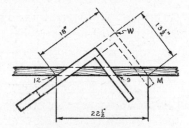

Fig. 72. Finding Difference in Lengths of Jack Rafters Which Are 18 Inches on Centers with a 9-Inch Rise per Foot Run

$12\ M$, which will then be from $18$ on the blade to $13\frac{1}{2}$ on the tongue, measured along the dressed edge of the lumber will be the difference in length of the jack rafters and will be $22\frac{1}{2}$ inches. This method can only be used when the spacing of the jack rafters center to center is not more than 18 inches, because if the spacing is more than this, it will be found that the tongue of the square is not long enough to reach from the heel to the dressed edge of the piece of lumber.

There is another way to get the lengths of jack rafters for hip and valley roofs *of equal pitch* by means of the steel square. On the face of the blade or body of those squares which have rafter tables, will be found a series of figures arranged in columns beneath the numbers in the inch line on the outside edge of the body or blade, see Fig. 37. The figures in the third line from the top in these columns give, in feet and inches, the difference in length of jacks— 16 inches on centers, and the figures in the fourth line from the top give the difference in the length of jacks, 2 feet on centers. These figures show, in feet and inches, the lengths of the first or shortest jack rafter (which is the same as the difference in length between the first and second jack, between the second and third jack and so on) for that roof slope of which the rise per foot run in inches is indicated by the number at the head of the column in the inch line on the outside edge of the body or blade of the square. Thus, to find the length of a jack rafter, multiply the number of feet and inches found in the rafter tables by the number indicating the position of the jack rafter with reference to the corner of the building or the point of intersection of the ridge boards; for example, multiply by 1 for the first or shortest jack, by 2 for the second or next longer jack, 3 for the third which is still longer, and so on. This gives the theoretical length from which must be subtracted half of the diagonal thickness of the hip or valley rafter and to which must be added an allowance for the tail or eave if the eaves project.

Suppose, for instance, that it is necessary to find the length of the third jack rafter in a roof slope which has a rise of 8 inches per foot of run of the common rafter, and that the spacing of the jacks center to center is 16 inches. On the outside edge of the face of the body or blade of a steel square having rafter tables find the figure *8* which indicates the rise per foot run of the roof. On the third line under this figure will be found the figures *"19-23"* (see Fig. 38) which means that the difference in length of the jack rafters or the length of the first and shortest jack rafter is **19.23** inches. So long as these figures are less than **24** inches they are given in inches; when they are greater than **24** inches, they are given in feet and inches; as, for instance, *2 10* meaning 2 feet 10 inches or *3 7¼*, meaning 3 feet and 7¼ inches). Because you want to get the length of the third shortest jack rafter, multiply the figure given, namely **19.23** inches by 3, and this gives us **57.69** inches, which is 4 feet and 9⅔ inches. This theoretical length of the jack rafter must be reduced by 1½ times half the thickness of the hip or valley rafter and must be increased by the necessary allowance for the tail where the jack rafter rests on and projects over the wall plate.

If the jack rafters were 24 inches apart center to center instead of 16 inches, it would be necessary to look in the *fourth* line under the figure 8, instead of the *third* line, to find the difference in length.

**Cripple Jack Rafters.** Fig. 73 shows the plan of a roof with hips and valleys. It shows some ordinary jack rafters such as those already described and at *a*, *b*, and *c* it shows some rafters which extend from the valley to the hip without touching either the wall plate or the ridge board. These are called cripple jacks. The best way to get the length of these jack rafters is to make a roof layout, like Fig. 73 but to a large scale, say to a scale of 1 inch to the foot.

If you were on the roof and able really to measure with a rule or tape the actual distance *on the slope* from the outside edge

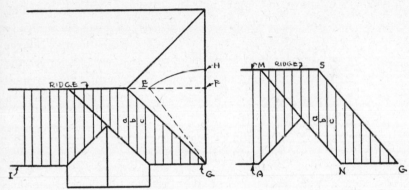

Fig. 73. Plan of Roof Which Has Cripple Jacks   Fig. 74. Plan of Roof to Show Actual Length of All Rafters

of the wall plate to the center of the ridge board, you would have the length of the common rafter, and you could measure the actual length of the jack rafters as well, including the cripple jacks. As this cannot be done, the next best thing is to make a plan to scale as shown in Fig. 73, looking straight down on the roof; then make another plan looking directly at the *slope* of the roof as shown in Fig. 74, which shows to scale the actual length of all the rafters.

To make this plan, imagine that the roof slope shown in plan at Fig. 73 is rotated downward around the wall line *IG* as an axis until it lies flat on the page before us as shown in Fig. 74. To do this lay out at *EF* the rise of the roof from top of wall plate to point where measuring line on common rafter intersects the center line of the ridge, the point *E* representing the upper end of the common rafter and the point *G* representing the lower end of the common rafter or the outside upper corner of the wall plate. Then with point *G* as a center rotate the line *EG* around point *G* until *EG* is in the position *GH*. Then make *MA* equal to *GH* and *EG* (the length of the common rafter) and lay out the roof slope as shown at Fig. 74. All the rafters are then shown to scale in their real length and this length can be scaled directly from the layout, if it is drawn to a scale of, for example, one inch to the foot.

Wherever the jack rafters rest against, or frame into, the wall plate or the ridge board, the cuts at the ends of the rafters will be similar to those for the common rafters, but where the jack rafters rest against, or frame into, the hip or the valley rafters, there will be side cuts as shown in Figs. 75 and 76. See also Fig. 30. To make

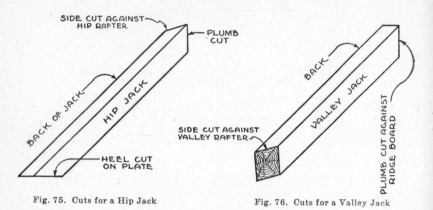

Fig. 75. Cuts for a Hip Jack          Fig. 76. Cuts for a Valley Jack

these side cuts, the steel square may be used in the same way as was explained and shown in Fig. 58 for getting the side cut for hip rafters.

Take a steel square having rafter tables and on the face of the body or blade of the square will be found a line of figures (the fifth from the top in each column) giving the figure to use on the body or blade of the square, together with 12 inches on the tongue, in order to lay out the side cut for jacks in roofs of equal pitch having various degrees of rise per foot of run. The rise per foot of run in inches is indicated in each case by the figure appearing in the inch line at the head of each column of figures in the rafter tables, see Fig. 37. For example, to find the side cut for the jack rafters in a roof with a rise of 8 inches per foot of run, look on the face of the body or blade of the steel square in the column of figures underneath the figure 8, and in the fifth line from the top will be found the figure *10*, see Fig. 38.

Select the piece of lumber from which the jack rafter is to be cut and dress one edge, which will be the top edge or back of the finished rafter. Then turn the piece of lumber so that you are looking straight down on this dressed top edge as shown in Fig. 77. Draw a line straight through the middle of the top edge lengthwise from end to end, as shown in the figure, to act as a measuring line. The point in which the top cut, or plumb cut, in Figs. 40 and 42, intersects the top edge of the jack rafter is actually on this measuring line

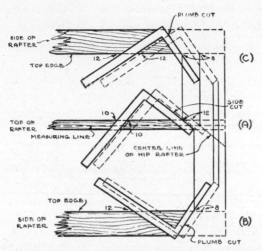

Fig. 77. Locating Side Cut and Plumb Cuts for a Jack Rafter

in the center of the top edge. Fig. 44 shows how allowance is made for the thickness of the ridge board in the case of common rafters, and a similar allowance must be made for the hip rafter when dealing with jacks.

Now take the steel square and apply it to the top edge, or back, of the jack rafter as shown in Fig. 77 in outline at (A), so that the 10-inch mark on the outside of the body or blade of the square (corresponding to the figure 10 just found in the rafter tables) comes on the edge of the back of the rafter, and the 12-inch mark on the tongue of the square also comes on this same edge of the back of the rafter, and so that the outside edge of the tongue of the square passes at the same time through the mark on the measuring line which indicates the point of intersection between the center line of the jack and the center line of the hip. Now mark along the outside edge of the tongue of the square, and this mark will represent the center line of the hip rafter. This line is parallel to the line of the side cut but does not represent the side cut itself because allowance must be made for the thickness of the hip rafter. This allowance is made by moving the square to the left a distance equal to 1½ times half the thickness of the hip rafter (as shown by the blacked-in outline of the square); a mark made along the outside edge of the tongue of the square in this new position will be the side cut (Fig. 75) on the top edge or back of the jack rafter.

In Fig. 77 (A), you are looking directly down on the top edge of the jack rafter, and the side-cut is shown marked off across this top edge. It is also necessary, however, to mark off the plumb cuts on the two *sides* of the jack rafter and these will not be opposite to each other, because the side cut on the top edge of the rafter shown in Fig. 77 (A) does not go square across the rafter at right angles to the sides. If you will imagine yourself to be looking straight down on the top edge of the jack rafter as shown in Fig. 77 (A) one side of the rafter will be *towards* you and the other side of the rafter will be *away* from you. The side which is *towards* you is shown in Fig. 77 (B), while the side which is *away* from you is shown upside down in Fig. 77 (C). Figs. 77 (B) and 77 (C) show the plumb cuts on the two sides of the jack rafter made along the outside edge of the *tongue* of the square. When the side cut is marked off across the top edge of the jack rafter as shown in Fig. 77 (A), it locates points on the top edge of each side of the rafter, and the plumb cuts are made by placing the outside edge of the tongue of the square against these points at a mark on the tongue of the square corresponding to *the rise per foot run* of the jack rafters (8 in this case). The square is then swung around so that the 12 on the outside edge of the blade of the square also comes on the top edge of the rafter and the plumb cut for each side of the rafter can be marked off along the outside edge of the tongue of the square.

To cut the roof boards to fit the line of the hip or valley in a roof of equal pitch, the square may be used as shown in Fig. 78. Taking the length of the common rafter per one foot run (given in the top line of figures in the rafter tables, Fig. 37), on the blade of the square and 12 inches on the tongue, apply the square to the roof board as shown in Fig. 78, and cut along the outside edge of the tongue. It will be noticed that this figure also illustrates another way of getting the bevel for jacks to fit against the hip rafter in roofs of equal pitch, by applying the square to the top edge or back of the jack, just as shown in Fig. 78 and as explained for roof boards, and then marking along the outside edge of the blade.

The ends of the roof boards resting on the hips or valleys would not be cut off square with the sides, but would have to be *mitered* in order to fit together. Fig. 79 shows how this miter can be cut. Using the steel square against the edge of the roof board with 15 on the blade (being the length of the common rafter per foot of run) and 9 on the tongue (being the rise of the roof slope per foot of run)

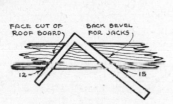

Fig. 78. Cuts for Roof Boards

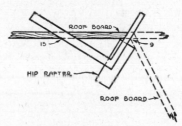

Fig. 79. Miter Cuts for Roof Boards
Resting on Hip or Valley Rafters

held on the line of intersection between the edge and the side of the board, the cut is made along the outside edge of the tongue of the square.

In Fig. 74 the hip rafter *SG* and the valley rafter *MN* would be shown in their true lengths to the scale used in laying out this figure, which might be 1 inch to the foot. Also, the angles or bevels which they make with the side of the ridge board would be shown in their true value by the angles which the lines *SG* and *MN* make with the ridge line *MS*, but these bevels can only be used when the hip rafters are *backed*, because Fig. 74 is a direct view, to scale, of the roof surface, and only the top edges of the hip rafters lie in the roof surface when the rafters are backed. Therefore, to use these bevels the steel square would have to be applied to the *backed* edges of the hip rafter and could not readily be applied to the valley rafter even if this rafter were backed. It has already been shown that the figures to be used on the steel square for this bevel when the hip rafter is left square on the back or top surfaces, as shown in Fig.

79, can be taken directly from the rafter tables on the steel square. See Fig. 58.

In case the square which the workman is using does not have the rafter tables, another method is given here. In order to make use of it, the tongue of the square used must be 18 inches long instead of the usual 16 inches. Such squares can be bought and they cost no more than the 16-inch squares. According to this second method, the square is applied to the square top edge of the hip rafter as shown in Fig. 80 and the dimensions to be used on tongue and blade are as follows:

The length of the run of the hip or valley per foot run of roof for roofs of equal pitch, which is 17, is used on the tongue and, on

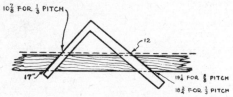

Fig. 80. Method of Finding Side Cut for Hip Rafter without Rafter Tables

the blade, the length of a hip or valley that will span over 17 inches run. A way to find this length is given later. A line drawn along the blade will give the bevel as shown in Fig. 80. Since the tongue of the ordinary square is only 16 inches long, it is only possible to do this with a special square which has an 18-inch tongue. For this reason the rafter tables are based on reducing the rafter length to 12 inches and the run in the same proportion as shown in Fig. 58.

For example, in Fig. 80, for a roof of ⅓ pitch the square is laid on the edge of the rafter in such a way that a line passes across the square from the 17-inch mark on the outside edge of the tongue to the 18¾-inch mark on the outside edge of the blade. Now suppose that a line parallel to this be passed through the 12-inch mark on the outside edge of the blade. It will be found that such a line would pass across the outside edge of the tongue of the square at the 10⅞-inch mark. It is therefore evident that instead of using 17 on the tongue and 18¾ on the blade for a ⅓ pitch roof, the same cut might be laid out by using 12 on the blade and 10⅞ on the tongue, or the square might be reversed with 12 on the tongue and 10⅞ on the

blade or body of the square, as is done in Fig. 58. The rafter tables
are based on this principle.

Fig. 81 contains all the bevels or cuts for a roof of equal pitch
such as has been considered in the preceding pages and with a rise
per foot run of 9 inches. This figure, if correctly understood, will
enable anyone to frame up any roof of this kind. In this figure

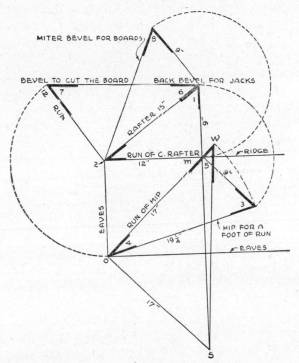

Fig. 81. Method of Finding Bevels for All Timbers in Roof of
Equal Pitch

it is shown that a 12-inch run and 9-inch rise will give bevels 1 and 2,
which are the plumb and heel cuts of the common rafters of a roof
rising 9 inches to the foot run. Therefore, by taking these figures
on the square, 9 inches on the tongue and 12 inches on the blade,
marking along the tongue will give the plumb cut and marking along
the blade will give the heel cut.

Bevels 3 and 4 are the plumb and heel cuts for the hip rafters.
The rise is 9 inches, and the run is the length of the run of the hip
rafter for one foot of run of the common rafter, which is 17 inches.

Therefore, take **17** inches on the blade and **9** inches on the tongue, and mark along the tongue for the plumb cut and along the blade for the heel cut.

Bevel *5,* which is to fit the hip or valley rafters against the ridge *when they are not backed,* is the angle opposite to the shortest side of a right triangle, *OWS,* one side of which, *OS,* is the *run* of the hip for one foot run of the roof (17 inches) and the other side of which, *OW,* is the *length* of the hip rafter for one foot run of the roof (19¼ inches).

These figures, therefore, taken on the square, 19¼ on the blade, and **17** on the tongue, will give the bevel by marking along the blade as shown in Fig. 80, where the square is shown applied to the hip timber with 19¼ on blade and **17** on tongue (only possible with 18-inch tongue), the blade showing the cut. For a square with 16-inch tongue, taking 18 on the blade and 16 on the tongue will give practically the same bevel.

Bevels *6* and *7* in Fig. 81 are shown formed of the length of the rafter for one foot of run, which is **15** inches, and the run of the rafter, which is **12** inches. These figures are applied on the square, as shown in Fig. 78, to a jack rafter timber; taking 15 on the blade and 12 on the tongue, marking along the blade will give the back bevel for the jack rafters, and marking along the tongue will give the face cut of roof boards to fit along the hip or valley.

It is also shown in Fig. 81 that by taking the length of rafter 15 inches on blade, and rise of roof 9 inches on tongue, bevel *8* will give the miter cut for the roof boards. See Fig. 79 also.

A similar figure can be drawn out for a roof of equal pitch having any other rise per foot of run instead of the 9-inch rise per foot of run provided for in Fig. 81; for instance, a roof with a rise per foot of run of 8 inches could be drawn. In this case distance *1-m* would be 8 instead of 9, and distance *2-1* would be 14½, as shown in Fig. 70, instead of 15. All the other bevels and lengths would change accordingly.

In Fig. 81 the angles are outlined in heavy black lines and this suggests the use of a tool called a *bevel,* the purpose of which is to transfer an angle or bevel from a drawing to the work or to adjust a fence to a steel square. Fig. 82 shows one of these tools and Fig. 83 illustrates its use.

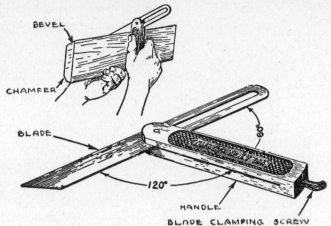

Fig. 82. An 8-Inch T Bevel

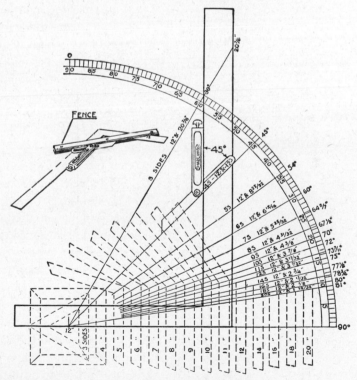

Fig. 83. Polygons and Their Miters

Set the bevels for these angles with the steel square. A fence of two strips of wood, shown above, will help to obtain a proper setting.

*Courtesy of The Stanley Rule and Level Plant, New Britain, Conn.*

Fig. 84 shows one half of a gable of a roof rising 9 inches to the foot run. The squares, placed as shown at the bottom and at the top, will give the plumb and heel cuts of the common rafter. The

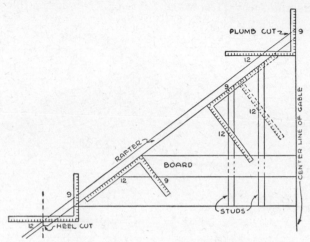

Fig. 84. Laying Out Timbers of One-Half Gable of ⅜-Pitch Roof

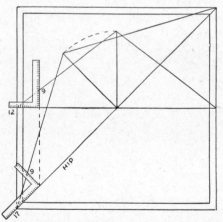

Fig. 85. Square Applied to Determine Relative Length of Run for Rafter and Hip

same figures on the square applied to the studding, marking along the tongue for the cut, will give the bevel to fit the studding against the rafter, and by marking along the blade the cut is obtained for the boards that run across the gable.

'Fig. 85 represents the plan of a square tower which is roofed over with a hipped roof in the shape of a pyramid. In this figure is shown the relative length of run for a common rafter and a hip, the rafter being 12 inches and the hip 17 inches. The reason, as shown in this diagram, why 17 is taken for the run of the hip, instead of

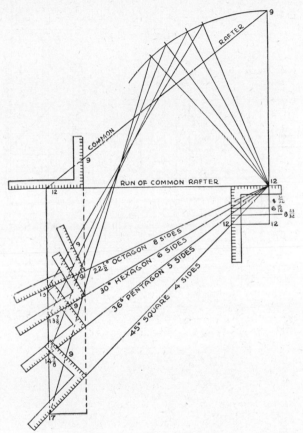

Fig. 86. Use of Square to Determine Length of Run for Hip Rafters on Corners Other Than 45°

12 as for the common rafter, is that the seats of the common rafter and hip do not run parallel with each other, but diverge in roofs of equal pitch at an angle of 45°; therefore, 17 inches taken on the run of the hip is equal to only 12 inches when taken on that of the common rafter, as shown by the dotted line from heel to heel of the two squares in Fig. 85.

In Fig. 86 is shown how other figures on the square may be found for corners that deviate from the 45°. It is shown that for a pentagon, which makes a 36° angle with the plate, the figure to be used on the square for run is 14⅞ inches; for a hexagon, which makes a 30° angle with the plate, the figure will be 13⅞ inches; and for an octagon, which makes an angle of 22½ degrees with the plate, the figure to use on the square for run of hip to correspond to the run of the common rafters, will be 13 inches. It will be observed that the height or rise in each case is 9 inches.

Note that in roofs of equal pitch the figure 12 on the blade,

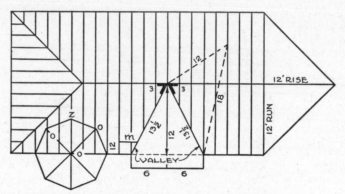

Fig. 87. Laying Out Timbers of Roof with Two Unequal Pitches

and whatever number of inches the roof rises to the foot run on the tongue, will give the plumb and heel cuts for the common rafter; and that by taking 17 on the blade instead of 12, and taking on the tongue the figure representing the rise of the roof to the foot run, the plumb and heel cuts are found for the hips and valleys.

By taking the length of the common rafter for one foot of run on blade, and the run 12 on tongue, marking along the blade will give the back bevel for the jack to fit the hip or valley, and marking along the tongue will give the bevel to cut the roof boards to fit the line of hip or valley upon the roof, see Fig. 78.

With this knowledge of what figures to use, and why they are used, it will be an easy matter for anyone to lay out all rafters for equal-pitch roofs.

In Fig. 87 is shown a plan of a roof with two unequal pitches. The main roof is shown by the diagram at the right to have a rise of 12 inches to the foot run. The front wing is shown to have a run of 6 feet and to rise 12 feet; it has thus a pitch of 24 inches to the foot run. Therefore 12 on blade of the square and 12 on tongue will give the plumb and heel cuts for the main roof, and by stepping 12 times along the rafter timber the length of the rafter is found, see Fig. 36. The figures on the square to find the heel and plumb cuts for the rafter in the front wing, will be 12 run and 24 rise, and by stepping 6 times (the number of feet in the run of the rafter), the length will be found over the run of 6 feet, and it will measure 13 feet 6 inches.

If, in place of stepping along the timber, the diagonal of 12 and 24 is multiplied by 6, the number of feet in the run, the length may be found to an even greater exactitude.

Many carpenters use this method of framing; and to those who have confidence in their ability to figure correctly, it is a saving of time, and, as before said, will result in a more accurate measurement; but the better and more scientific method of framing is to work to a scale of one inch, as has already been explained.

According to that method, the diagonal of a foot of run, and the number of inches of rise per foot of run, measured to a scale, will give the exact length. For example, the main roof in Fig. 87 is rising 12 inches to a foot of run. The diagonal of 12 and 12 is 17 inches, which, considered at a scale of 1 inch to a foot, will give 17 feet, and this will be the exact length of the rafter for a roof rising 12 inches to the foot run and having a run of 12 feet.

The length of the rafter for the front wing, which has a run of 6 feet and a rise of 12 feet, may be obtained by placing the rule as shown in Fig. 88 from 6 on blade to 12 on tongue, which will give a length of 13½ inches. If the scale be considered as one inch to a foot, this will equal 13 feet 6 inches, which will be the exact length of a common rafter rising 24 inches to the foot run and having a run of 6 feet.

It will be observed that the plan lines of the valleys in Fig. 87 in respect to one another deviate from forming a right angle. In equal-pitch roofs the plan lines are always at right angles to each other, and therefore the diagonal of 12 and 12, which is 17 inches, will be the relative foot run of valleys and hips in equal-pitch roofs.

In Fig. 87 is shown how to find the figures to use on the square for valleys and hips when deviating from the right angle. A line is drawn at a distance of 12 inches from the plate and parallel to it, cutting the valley in *m* as shown. The part of the valley from *m* to the plate will measure 13½ inches, which is the figure that is to be used on the square to obtain the length and cuts of the valleys.

It will be observed that this equals the length of the common rafter as found by the square and rule in Fig. 88. In that figure is shown 12 on blade and 6 on tongue. The 12 here represents the rise, and the 6 the run of the front roof. If the 12 be taken to represent the run of the main roof, and the 6 to represent the run of the front

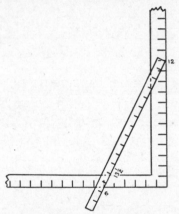

Fig. 88. Finding Length of Rafter for
Front Wing in Roof Shown in Fig. 87

roof, then the diagonal 13½ will indicate the length of the run of the valley for 12 feet of run of common rafter in the main roof, and therefore for one foot of run it will be 13½ inches. Now, by taking 13½ on the blade for run, and 12 inches on the tongue for rise, and stepping along the valley rafter timber 12 times, the length of the valley will be found. The blade will give the heel cut, and the tongue the plumb cut.

In Fig. 89 is shown the slope of the roof projected into the horizontal plane. By drawing a figure based on a scale of one inch to one foot, all the timbers on the slope of the roof can be measured. Bevel 2, shown in this figure, is to fit the valleys against the ridge. By drawing a line from *w* square to the seat of the valley to *m*, making *w 2* equal in length to the length of the valley, as shown, and by connecting *2* and *m*, the bevel at *2* is found, which will fit the valleys against the ridge, as shown at *3* and *3* in Fig. 87.

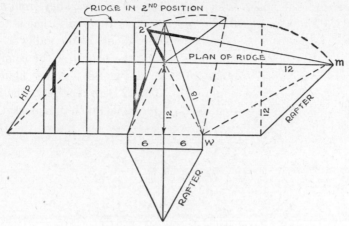

Fig. 89. Laying Out Timbers of Roof Shown in Fig. 87, by Projecting Slope of Roof into Horizontal Plane

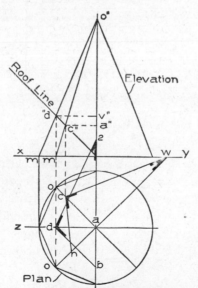

Fig. 90. Method of Finding Length and Cuts of Octagon Hips Intersecting a Roof

In Fig. 90, is shown how to find the length and cuts of octagon hips intersecting a roof. In Fig. 87, half the plan of the octagon is shown to be inside of the plate, and the hips, *o, z, o* intersect the slope of the roof. In Fig. 90, the lines below $xy$ are the plan lines; the lines above $xy$ show the elevation. From $z$, $o$, $o$, in the plan, draw lines to $xy$, as shown from $o$ to $m$ and from $z$ to $m$; from $m$ and $m$, draw the elevation lines to the apex $o''$, intersecting the line of the roof in $d''$ and $c''$. From $d''$ and $c''$, draw the lines $d''v''$ and $c''a''$ parallel to $xy$; from $c''$, drop a line to intersect the plan line $ao$ in $c$. Make $aw$ equal in length to $a''o''$ of the elevation, and connect $wc$; measure from $w$ to $n$ the full height of the octagon as shown from $xy$ to the apex $o''$; and connect $cn$. The length from $w$ to $c$ is that of the two

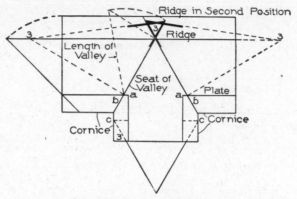

Fig. 91.  Showing How Cornice Affects Valleys and Plates in Roof with Unequal Pitches

hips shown at $o$, $o$ in Fig. 87, both being equal hips intersecting the roof at an equal distance from the plate. The bevel at $w$ is the top bevel, and the bevel at $c$ will fit the roof.

Again, drop a line from $d''$ to intersect the plan line $az$ in $d$. Make $a$ $2$ equal to $v''o''$ in the elevation, and connect $2d$. Measure from $2$ to $b$ the full height of the tower as shown from $xy$ to the apex $o''$ in the elevation, and connect $db$. The length $2d$ represents the length of the hip $z$ shown in Fig. 87; the bevel at $2$ is that of the top; and the bevel at $d$, the one that will fit the foot of the hip to the intersecting roof.

When a cornice of any considerable width runs around a roof of this kind, it affects the plates and the angle of the valleys as shown in Fig. 91. In this figure are shown the same valleys as in Fig. 87; but, owing to the width of the cornice, the foot of each has been moved the distance $ab$ along the plate of the main roof.

The bevels shown at *3,* Fig. 91, are to fit the valleys against the ridge. This is done because it is necessary for the valleys to intersect the corners *C, C,* of the cornice at these points, which are the points where the lines representing the outside edges of the two cornices intersect. It will be observed from an inspection of Fig. 91 that the cornice along the eaves of the narrow ell projects outside of, or overhangs, the wall line the same distance as does the cornice along the eaves of the wider main roof, or, in other words, that the two cornices are of the same *width* despite the fact that the ell roof

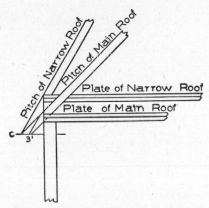

Fig. 92. Showing Relative Position of
Plates in Roof with Two Unequal Pitches

is much narrower than the main roof. If the cornice along the eaves of the narrow ell roof could have been made with a narrower overhang than the cornice along the eaves of the wider main roof, the outside lines of the two cornices might have intersected in points *C, C,* so located that the line representing the "Seat of Valley" would have passed through point *a* which is at the intersection of the lines representing the outside edges of the two wall plates. However, it is often desirable to give the two cornices, one at the eaves of the narrow ell and the other at the eaves of the wider main roof, the same overhang, as shown in section in Fig. 92 and to have the lines of the eaves, marked *C* in Fig. 92, at the same level in both cases.

Assume that *C,* in Fig. 92, is a point on the measuring line of the rafters in the wide main roof and also in the narrow ell roof, and that the two ridges are at the same height, that is, at the same distance from point *C* in a vertical direction. Owing to the fact that the horizontal distance from point $C$ to the ridge line is much greater in the case of the main roof than it is in the case of the narrower ell roof, the pitch of the narrow roof will be much steeper than the pitch of the main roof, as shown in Fig. 92, both slopes starting from the same level at point $C$, 3 feet out from the line of the outside edge of the wall plate. The slope or pitch of the rafters is determined in each case by the vertical distance from the eave line at $C$, Fig. 92, up to the ridge line, and by the width of the roof from point $C$ right

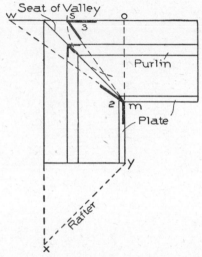

Fig. 93.  Method of Finding Bevels for Purlins in Equal-Pitched Roofs

across to a similar point on the other side of the roof. Since the width is different in each roof, the narrower roof has the steeper pitch.

In order that the rafters may, in each case, rest on the wall plates, the plate for the narrower ell roof must be placed at a higher level than the plate for the wider main roof, as shown in Fig. 92 where the plate for the narrow roof is shown to be much higher than the plate for the main roof.

In Fig. 93 is shown a very simple method of finding the bevels for purlins in equal-pitch roofs. Draw the plan of the corner as shown, and a line from $m$ to $o$; measure from $o$ the length $xy$, representing the common rafter, to $w$; from $w$ draw a line to $m$; the bevel shown at $2$ will fit the top face of the purlin. Again, from $o$, describe an arc to cut the seat of the valley, and continue same around to *S;* connect *Sm;* the bevel at *3* will be the side bevel.

This method is strictly correct only for roofs where the *rise* of the rafters is equal to the *run* as indicated in the lower part of Fig. 93, but it will give results sufficiently accurate for other roofs, provided that they are neither very steep nor very flat.

The graceful beauty of this stairway represents a mathematical accomplishment of the first order. From the job of working out the over-all design to that of laying out treads and risers, the steel square is the craftsman's major tool.

# CHAPTER IV

## OTHER USES OF THE STEEL SQUARE

**Use of the Octagon Scale.** When describing the steel square in the first part of the book, reference was made to the Octagon Scale or Eight Square, which is found on the face of the tongue of the

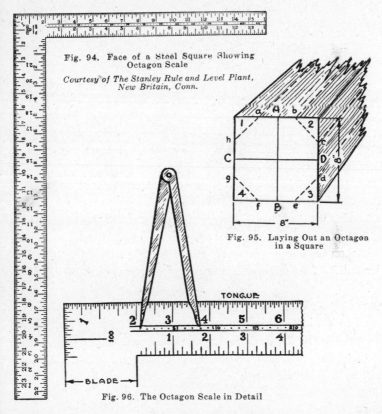

Fig. 94. Face of a Steel Square Showing Octagon Scale

*Courtesy of The Stanley Rule and Level Plant, New Britain, Conn.*

Fig. 95. Laying Out an Octagon in a Square

Fig. 96. The Octagon Scale in Detail

square. As was there described, this consists of a series of divisions in the shape of dots marked off along the middle of the tongue of the square, starting nearly under the 2-inch mark on the outside edge near the heel and continuing nearly to the end of the 16-inch long tongue where the figure *65* appears, as shown in Fig. 94. There are 65 of these dots and every fifth one is numbered, thus: 5, 10, 15, 20, 25, 30, 35, etc., up to *65*. This scale is used for the purpose of laying out octagons or figures with 8 equal sides, sometimes called *eight-squares* just as a figure with 4 equal sides is called a square.

Occasionally it becomes necessary for a carpenter or joiner or a cabinet maker to transform a square stick of timber into an eight-sided one, such, for example, as an octagonal newel post for a stair. To do this, it is necessary to lay out an eight-square or octagon on the end of a square stick of stuff. The method is as follows:

Cut the end of the stick square with the sides. The end will be a square; assume it to be 8 inches on each side. Find the center of each side as shown at *A, B, C* and *D* in Fig. 95, and draw lines *AB* and *CD*. With dividers or a rule, measure off on the octagon or eight-square scale on the tongue of the square the length of 8 spaces, since the timber is 8 inches square. See Fig. 96. If the timber were 10 inches square, the length of 10 spaces would be measured off; if it were 12 inches square, 12 spaces, and so on.

Having measured off the length of 8 spaces on the octagon scale (a distance of a little less than 2 inches), apply this measurement to each side of the square timber on both sides of the center points, *A, B, C* and *D*, as *Aa, Ab, Bf, Be, Ch, Cg, Dc* and *Dd*, Fig. 95. Now, joining the points *ah, bc, de* and *fg*, will outline a figure having 8 equal sides on the end of the stick, and it can then be shaped to this form by cutting off the solid triangular pieces from each of the four corners.

**Use of Steel Square for Beveling.** In order to save material and labor and to avoid trouble due to "checking" of the timber, which causes cracks in the finished work, newel posts and other such members are seldom cut out of the solid wood as described above, but are formed out of comparatively thin boards and are hollow inside. The sides must then be mitered together at the corners and glued or splined to each other to form the post which may have 4, 5, 6 or 8 sides. In order to make the sides fit properly together, their edges must be cut to the right bevel or miter and this is done with the aid of the steel square and a *bevel*, shown in Fig. 82. The bevel can be set to the proper angle between the blade and the handle and then clamped tight by means of the clamping screw.

The bevel can be set to the proper angle by means of the steel square as shown in Fig. 83. Take the square so that you are looking at the face. Lay it on a wide, smooth board or piece of paper with the blade pointing *away* from you and with the tongue extending to the left; mark at the 12-inch division on the outside of the tongue, which will be towards you. Mark along the outside edge of the blade at a division mark depending on the number of sides to the figure. For example, if you are dealing with a 6-sided post, find the sloping line marked *6S* in Fig. 83, and note that on this line it says, *12* inches and $6^{15}\!/_{16}$ inches. Therefore for a 6-sided figure make a mark opposite $6^{15}\!/_{16}$ inches on the outside edge of the blade of the square. Draw a line joining the 12-inch mark on the tongue and $6^{15}\!/_{16}$-inch mark on the blade. Replace the square on the paper just as before and set the bevel with the handle against the inside edge of the blade and the blade along the mark just made. Then clamp the bevel and the required angle (60°) will be given between the handle and the blade as shown in Fig. 82. If a 5-sided figure is required, use the dimension $8^{25}\!/_{32}$ as given on the sloping line marked *5S*, in Fig. 83 ($8^{24}\!/_{32}$ inches or $8^{12}\!/_{16}$ inches or $8\frac{3}{4}$ inches is very nearly correct). Fig. 83 gives the angles and bevels and required dimensions for figures of different numbers of sides from 3 to 20 sides.

Fig. 83 also illustrates how to use a fence such as was described in connection with Fig. 9 to get the setting for the bevel instead of drawing the line on paper.

**Use of Steel Square in Building a Ladder Stair.** Fig. 97 shows the application of the steel square to laying out the strings for a very steep stair with treads but without risers, of a type called a ship's ladder stair, for access to a cellar or attic. At the left of the figure is shown a rod about 1 inch by 1 inch, cut to the exact length equal to the distance from one floor level to the other, in this case 7 feet and 6 inches. The well hole in the upper floor is 5 feet long, face to face of trimmer joists, so the run of the stair will be taken as 4 feet and 9 inches to allow easy passage up and down the ladder stair. For a steep ladder stair, the height from tread to tread should be 9 or 10 inches. This would give 10 risers at 9 inches each, or 9 risers at 10 inches each. Let us assume that there will be 10 risers and mark off the pole in 10 equal divisions.

The strings can be set up temporarily in position with the top edge dressed and the ends cut to fit against the floor at the bottom and the trimmer joist at the top.  Fit the fence to the steel square so that the tongue of the square is level and the blade plumb, while the fence fits snug against the top edge of the string.  Clamp the fence tight to the square in this position.  Set up the pole plumb

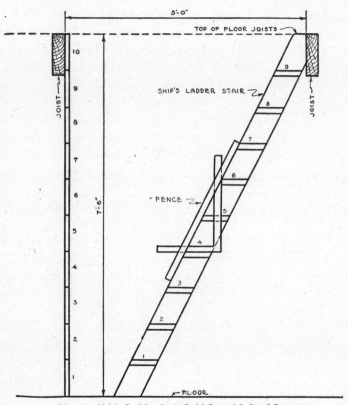

Fig. 97.  Ship's Ladder Stair Laid Out with Steel Square

in the well hole as shown and, by leveling across, mark off the position of the treads on the edge of the string of the ladder stair.  Then by moving the square along from one mark to the next, the position of the treads can be marked out on the string.

The thickness of the treads can be marked off as indicated at *1, 2, 3, 4,* etc., and if desired, channels can be cut into the sides of the strings at these points to receive the ends of the treads.  This is called *housing* the treads into the strings.  There are no risers, because in such very steep ladder stairs the toe of the shoe resting on any tread must be able to project underneath the face of the tread just above the one on which the foot rests. Risers would prevent this, so they must be omitted.

**Use of Steel Square in Making Braces.** The use of the steel square in the laying out and cutting of braces as illustrated and explained in connection with Figs. 10 to 14 applies only to braces with a vertical surface at the foot of the brace and a horizontal surface at the upper end of the brace as shown in these figures. The vertical surface may be the side of a corner post in a building frame and the horizontal surface may be the under side of a girt in the same frame.

There is another sort of brace which requires a somewhat different treatment when it is being laid out with the help of the steel square. This variety of brace has both ends resting against, or framed into, vertical surfaces of limited height. Examples of this sort of brace on a small scale are the cross bridgings between joists in a floor frame shown at A and B in Fig. 98. In this figure, C is the bridging.

On a larger scale such braces occur in the framing for braced towers or similar structures as illustrated in Fig. 99 where D is the brace. Short braces spanning between two vertical surfaces less than 24 inches apart, such as the floor joists A and B in Fig. 98, can be laid out and cut with one application of the steel square as shown in the figure, but larger braces where the distance between the two vertical surfaces (the run of the brace) is several times 24 inches, require different and somewhat more complicated treatment.

Let us consider the case of the bridging between the floor joists Fig. 98. Suppose the joists are 2x10 inches at 16 inches on center. These joists will actually be $1\frac{5}{8}$ inches thick by $9\frac{1}{2}$ inches deep. The rise of the piece of bridging, C in Fig. 98, will be the depth of the joist B, $9\frac{1}{2}$ inches, and the run of the piece of bridging will be the distance face to face of joists, which is 16 inches less $1\frac{5}{8}$ inches, or $14\frac{3}{8}$ inches. Now take the rise, $9\frac{1}{2}$ inches, on the outside edge of the tongue of the steel square and the run, $14\frac{3}{8}$ inches, on the outside edge of the blade or body as shown in Fig. 98. Notice that the $9\frac{1}{2}$-inch mark on the tongue of the square is held against the *upper* side of the piece of bridging C, while the $14\frac{3}{8}$-inch mark on the body of the square is held against the *lower* side of the bridging. It will be seen that this procedure is different from that followed in the case of the other braces shown in Figs. 10 to 14 and the rafters, where in every case one edge or side only of the brace or rafter, or in some cases a line parallel to the top edge of a rafter, was used as the measuring line. A cut made along the tongue of the square, Fig. 98,

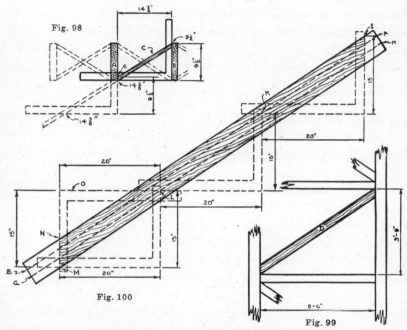

Figs. 98 to 100. Braces Laid Out with Steel Square

will fit the bridging against the right-hand joist *B*, and by moving the square along to the left, as shown by the dotted lines, so that the outside edge of the tongue comes at the mark made on the *lower* side of the bridging by the body of the square in its first position, and so that the 14⅜-inch mark on the body is again on the lower side of the stuff further to the left, while the 9½-inch mark is on the upper side of the bridging at *e*, Fig. 98, then a cut along the tongue will fit the bridging against the left-hand joist *A*. A length of stuff can in this way be cut up into a number of short lengths of bridging which will fit correctly between the joists, but it must be remembered that the mark on the outside edge of the tongue of the square corresponding to the depth of the joists must be kept at the *upper* side of the stuff while the point on the body of the square corresponding to the *clear spacing* face to face of the joists must be kept on the *lower* side of the stuff.

Let us now consider the longer brace for the braced tower as shown in Fig. 99. The run is 5 feet or 60 inches and the rise is 3 feet 9 inches, or 45 inches. If it were to be had and if it could be handled, a huge square could be used just as was done in the case of the bridging, but as such a square would be 3 times bigger than

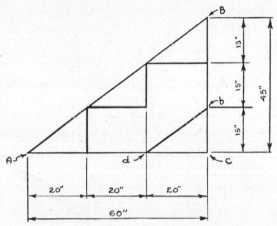

Fig. 101. Small Triangle Similar to Larger Triangle

the ordinary steel square and require a tongue 48 inches long, you apply the law of similar triangles, just as was done in the case of the rafter in Fig. 36, that the angle of the hypotenuse (the long side opposite to the 90-degree angle) of any right-angled triangle (such as *AB* in the triangle *ABC* in Fig. 101) is the same as the angle of the hypotenuse of any smaller right-angled triangle, provided that the smaller triangle is similar to the bigger one, that is, that all three of its sides are obtained by dividing the length of the similar sides of the larger triangle by the same number.

In Fig. 101 small triangle *dbC* is similar to the large triangle *ABC*. In the case of the brace shown in Fig. 99, to get a similar triangle within the limits of the steel square, divide the rise of 45 inches by 3, giving 15 inches as the figure to use on the tongue; then divide the run of 60 inches by 3, giving 20 inches as the figure to use on the blade or body of the square. The common divisor in this case is 3. Then if the steel square with 15 on the tongue and 20 on the body is applied to the piece of lumber for the brace three times in the proper way, the correct length is obtained and also the cuts at the upper and lower ends of the brace.

Just as in the case of the piece of bridging, the mark 15 on the tongue and 20 on the body of the square must *not* both be applied to the same edge of the piece of lumber as would be done in the case of a rafter. It is done in the case of a rafter because the rafter fits against a vertical surface at the top end, the ridge board, and against a horizontal surface at the bottom end, the wall plate.

This is different from the braced-tower brace being considered, which must fit between two vertical surfaces—one at each end. In this case make marks lengthwise on the side of the brace with pencil or scratch awl, dividing it into 3 parts—the same number as the required applications of the steel square to the brace. This is illustrated in Fig. 100 by the lines $EF$ and $GH$ which are parallel to the top and bottom edges of the piece. Fig. 100 is a larger view of the piece of lumber from which the brace $D$ of Fig. 99 is to be cut. This piece of lumber is about 7 feet long.

For the first position of the steel square at the upper *right* hand of Fig. 100, place the 15-inch mark of the outside edge of the tongue of the square on the upper edge of the lumber at $I$ and the 20-inch mark of the body or blade of the square on the upper division line $EF$ at $K$. A line at $I$ drawn along the outside edge of the tongue will be the plumb line, sometimes called the *miter line*, at the upper end of the brace. The point $K$ should now be located and marked on the upper division line $EF$ where the outside edge of the body of the square crosses this line at the 20-inch mark on the square.

For the second position of the square, place the 15-inch mark of the tongue at the point $K$ on the *upper* division line $EF$ where the body of the square in its first position crossed this line. Move the square about, still keeping the 15-inch mark on the tongue steadily at point $K$, until the 20-inch mark on the outside edge of the body or blade rests on the *lower* division line $GH$, as shown at $L$ in Fig. 100.

For the third position of the square, place the 15-inch mark on the tongue at the point $L$ on the *lower* division line $GH$ where the body of the square in its second position crossed this line.  Now move the square about, still keeping the 15-inch mark on the tongue at the point $L$, until the 20-inch mark on the body rests on the lower edge of the piece of lumber at $M$, as shown in Fig. 100.  A line $MN$ drawn through point $M$ parallel to the plumb line at $I$ will give the cut to fit the brace against the vertical post at the foot of the brace.  This line $MN$ can be drawn on the side of the brace by reversing the square as shown at $O$, Fig. 100, placing the 20-inch mark on the outside edge of the blade at $L$ and the 15-inch mark on the outside edge of the tongue at $M$.

Every finished house is literally the product of thousands of calculations that require the intelligent use of the steel square. Perfection is the product of the expert craftsman, not of the hammer-and-saw mechanic.

# CHAPTER V

## ILLUSTRATIVE PROBLEM

To make clear some of the many ways in which the steel square can be used by a carpenter in building a house, the construction of a dwelling will be followed through from start to finish; and the more usual applications of the square to the work will be illustrated and explained. For this purpose, a country home has been chosen, and the plans and elevations are reproduced in Figs. 102 to 110. These include: the basement plan (Fig. 102), the first-floor plan (Fig. 103), the second-floor plan (Fig. 104), a roof framing plan (Fig. 105), the north elevation (Fig. 106), the south elevation (Fig. 107), the east elevation (Fig. 108), the west elevation (Fig. 109), and a typical section (Fig. 110).

**Laying Out Wall Lines.** Try to imagine that a site has been chosen and that you as carpenter–builder have arrived on the ground with some helpers ready to start on the work.

The first job is to stake out the building, put up batter boards, and run lines so that the excavation can be made and the footings and foundation walls built. If these footings and walls are to be of concrete, the necessary wood forms must be constructed to the proper lines and levels so the concrete can be poured. If the foundation walls are to be built of concrete blocks or, which is less likely, of rubble masonry, lines must be run to serve as guides for the workmen who will build them.

It is comparatively easy to establish the first line which is usually the line of the front wall of the house and parallel to the highway, road, or street which the house will face.

After this line has been established, the side lines must be run at exactly right angles to it so the corners of the building will be *square*. This is very important, since any mistake which results in the foundation walls being built with the corners not truly square will be the cause of grief later on.

As nearly all carpenters know, the most common method of making sure that the lines at the corners are truly square is to use

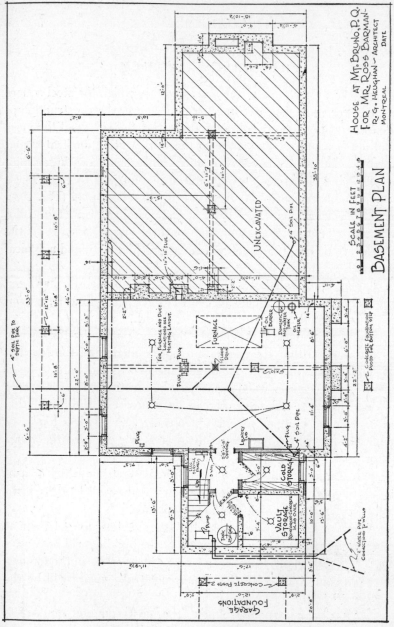

Fig. 102. Basement Plan

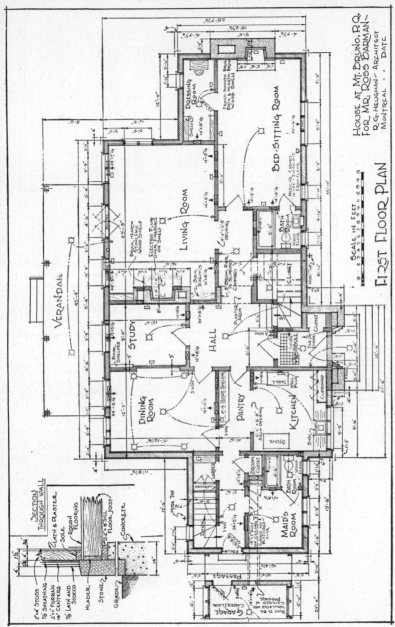

Fig. 103, First-Floor Plan

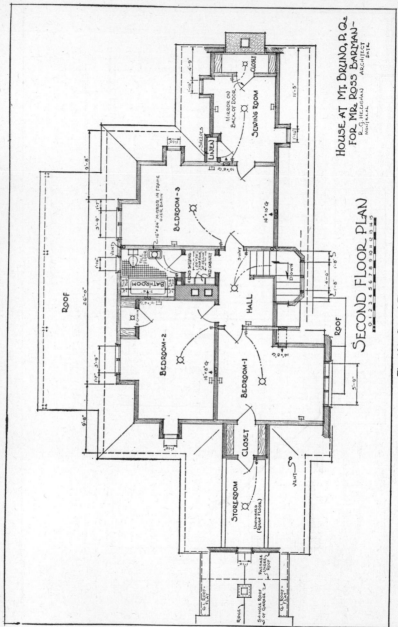

Fig. 104. Second-Floor Plan

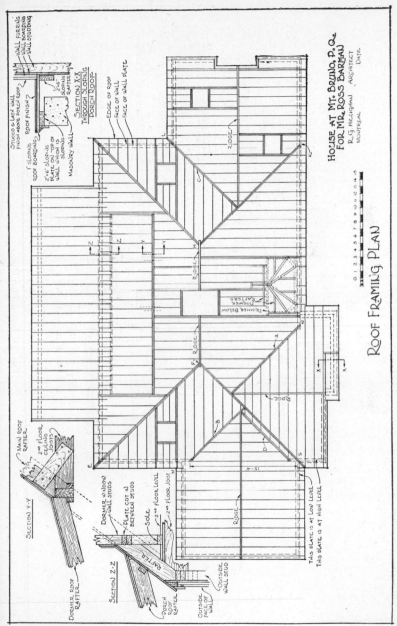

Fig. 105. Roof Framing Plan

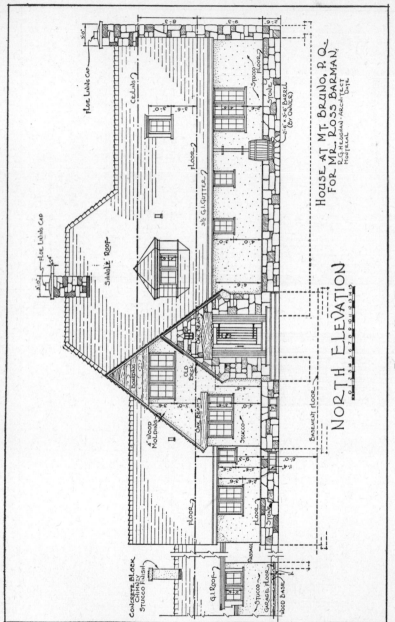

Fig. 106, North Elevation

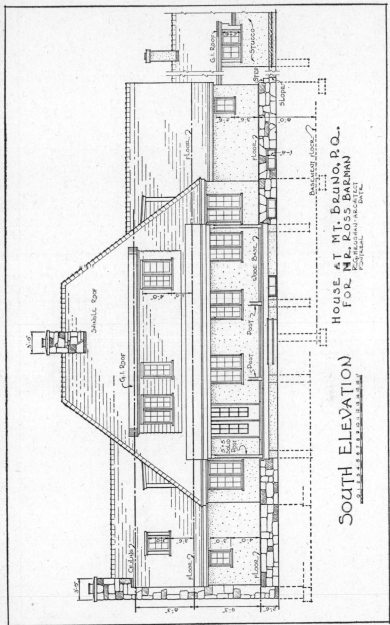

SOUTH ELEVATION

HOUSE AT MT. BRUNO, P.Q.
FOR MR. ROSS BARMAN
R.G. HEUGHAN ARCHITECT
MONTREAL

Fig. 107. South Elevation

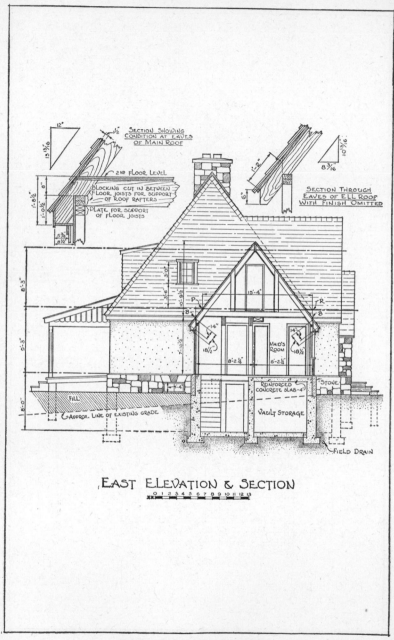

EAST ELEVATION & SECTION

0 1 2 3 4 5 6 7 8 9 10 11 12 13

Fig. 108. East Elevation and Section

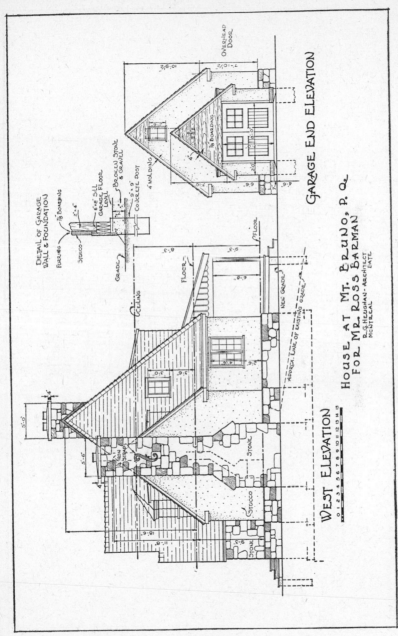

Fig. 109. West Elevation and Garage-End Elevation

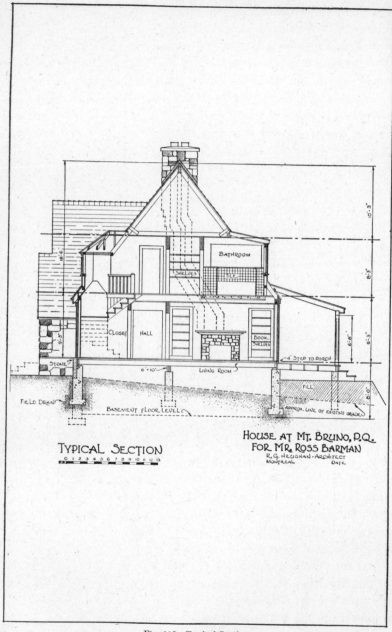

TYPICAL SECTION

0 1 2 3 4 5 6 7 8 9 10 11 12 13

HOUSE AT MT. BRUNO, P.Q.
FOR MR. ROSS BARMAN
R. G. HEUGHAN - ARCHITECT
MONTREAL          DATE

Fig. 110. Typical Section

a straight pole, 10 feet long; see Fig. 111. A marker, which may be a piece of cord knotted around the line, is fixed in place on the *front* wall line exactly 6 feet from the corner where the two lines meet. Another marker is placed on the *side* wall line exactly 8 feet from the corner where the two lines meet. The 10-foot pole, held level, is extended diagonally across the corner between the front wall line and the side wall line with one end on the marker on the front wall line, as shown at point $B$ in Fig. 111, and with the other end of the pole moved around until it touches the side wall line.

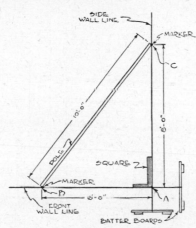

Fig. 111. Using 6–8–10 Rule for Squaring
Corner of Building

If the end of the 10-foot pole touches the side wall line exactly at the marker on this line (point $C$, Fig. 111, 8 feet from the meeting point of the two lines at the corner) while the other end of the pole remains at the marker on the front wall line (point $B$, Fig. 111, 6 feet from the meeting point of the two lines at the corner), then the two lines are exactly at right angles to each other; and the corner, if built to these lines, will be truly square.

If the pole does not touch the side wall line just at the marker on this line, then the two lines are not square with each other; the side wall line must be moved slightly one way or the other until the marker on it exactly coincides with the end of the 10-foot pole.

This method of squaring two lines at the corner of a building is sometimes called the 6–8–10 rule. Many tradesmen know of the rule and use it, but do not know why it is accurate. The steel square supplies the reason for this and proves the rule. It is known that the two outside edges of a steel square meeting at the heel of the square must be truly at right angles to each other. Now take a square and lay it down flat on a large piece of brown paper. Mark one line on the paper along the outside edge of the blade of the square (line *A–D*, Fig. 112) and another line on the paper along the outside edge of the tongue of the square (line *A–E*, Fig. 112). The two lines will meet at the heel of the square (point *A*, Fig. 112)

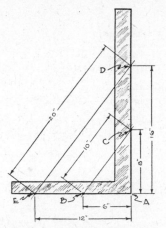

Fig. 112. Proof of 6–8–10 Rule by
Steel Square

and must be exactly at right angles to each other. Make a mark on the line drawn along the outside edge of the blade just 8 inches from the heel of the square (point *C*, Fig. 112) and make another mark (point *B*, Fig. 112) on the line drawn along the tongue of the square just 6 inches from the heel.

Now, with a foot rule, measure the distance diagonally across between these two points (*B–C*, Fig. 112), and you will find it is exactly 10 inches. Again, make a mark on the line drawn along the outside edge of the blade of the square just 16 inches (twice 8 inches) from the heel (point *D*, Fig. 112) and make a mark on the line drawn along the tongue of the square just 12 inches (twice 6 inches) from the heel (point *E*, Fig. 112). The distance diagonally across between these points (*D–E*, Fig. 112) is by measurement exactly 20 inches (twice 10 inches).

Thus, each of the distances can be multiplied by 2 and the same relation between them will exist. If they can be multiplied by 2 without disturbing the relationship between them, then they can be multiplied by 12 and the same relationship will exist. If each of the distances is multiplied by 12, the 6 inches become 6 feet, the 8 inches become 8 feet, and the 10 inches become 10 feet. Therefore, it is clear that if two lines are apparently at right angles or square with each other, meeting at a point of intersection such as point *A* in Fig. 111, and if you measure off 8 feet from the point of intersection along one line and 6 feet along the other line, then the two lines will be truly square with each other when the diagonal distance between the two points is exactly 10 feet, because all of these distances in feet are just exactly 12 times similar distances in inches which can be actually measured off on the steel square.

**Bridging for Floor Joists.** After the foundation walls of a house are poured and have set (if of concrete) or have been built (if of concrete blocks), the wood sills are placed on top of the foundation walls, and the floor joists for the first-floor framing are set with their ends resting on the wood sills. It is almost always necessary to put bridging between the floor joists to stiffen and strengthen the floor framing. Such bridging usually consists of pieces cut from 2- by 3-inch scantling (or 1- by 3-inch stuff) set in between the floor joists in pairs; the two pieces of each pair cross each other diagonally so that one end of each piece of bridging rests against the side of a floor joist near its top edge, while the other end rests against the next floor joist near its bottom edge, as shown in Fig. 98. The bridging will be more useful for stiffening the joists and will make a much stronger floor framing if the pieces forming the bridging have their ends cut to fit snugly against the sides of the floor joists as illustrated in Fig. 98, which also shows how the steel square can be used for this purpose.

In the case of this particular house, some of the floor joists are 2 by 8 (actually 1⅝ inches by 7½ inches), some are 2 by 10 (actually 1⅝ inches by 9½ inches), and some are 2 by 12 (actually 1⅝ inches by 11½ inches). Assume that the bridging is of 2- by 3-inch stuff. A quantity of pieces must be cut to fit between the 2- by 8-inch joists which are spaced 16 inches apart center-to-center. These joists will be really 7½ inches deep, and, since they are actually only 1⅝ inches thick, the clear spaces between them from face-to-face of joists will be 14⅜ inches.

To cut the pieces of bridging to fit in between the faces of the joists, take a long piece of 2- by 3-inch stuff. Lay the steel square flat against the 1⅝-inch edge of the stuff, as shown in Fig. 113, in such a way that the 7½-inch division on the outside edge of the tongue of the square will rest on the line of intersection between the side of the stuff farthest from the heel of the square and the edge of the stuff (at *B*, Fig. 113); at the same time, the 14⅜-inch division on the outside edge of the blade or body of the steel square should rest on the line of intersection between the *other* side of the stuff and the edge against which the square lies (at *C*, Fig. 113).

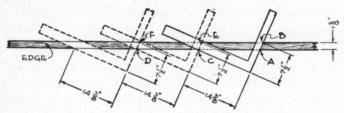

Fig. 113.  Laying Out Bridging with Steel Square

Now mark off a line across the edge of the stuff along the outside edge of the tongue of the square (*A–B* in Fig. 113), and make a mark at the point where the 14⅜-inch division on the outside edge of the blade of the square crosses the line of intersection between the edge of the stuff and the side of the stuff *nearest* to the heel of the square (point *C* in Fig. 113). Next, move the square to the left along the edge of the piece of stuff (against which it is still kept lying flat) until the outside edge of the tongue of the square crosses the edge of the stuff at the mark where the 14⅜-inch division on the outside edge of the blade was before the square was moved (point *C*, Fig. 113). The 7½-inch division on the outside edge of the tongue of the square still rests on the line of intersection between the edge of the stuff and the side of the stuff *farthest* from the heel of the square the same as before but in a different place (point *E*, Fig. 113), and the 14⅜-inch division on the outside edge of the blade of the square still rests on the line of intersection between the edge of the stuff and the side of the stuff *nearest* to the heel of the square the same as before but in a different place (point *D*, Fig. 113). This position of the steel square is shown by dotted lines in Fig. 113.

Make a mark (*C–E*, Fig. 113) across the edge of the stuff along the outside edge of the tongue of the square when the square is in its new position as shown by dotted lines in Fig. 113. Again mark the point where the 14⅜-inch division on the outside edge of the blade or body of the square crosses the line of intersection of the edge of the stuff with the side of the stuff nearest to the heel of the square (point *D*, Fig. 113). Then move the square the same as before and make another mark (*D–F*, Fig. 113) across the edge of the stuff along the outside edge of the tongue of the square. Repeat this operation as many times as the length of the piece of scantling will allow.

A series of marks or lines now extend diagonally across the edge of the stuff. From each end of each of these lines (that is, from points *A*, *B*, *C*, *E*, *D*, and *F*, etc., Fig. 113) mark off lines extending across the two sides of the stuff square with the edge of the stuff. This can be done with the aid of the steel square, as shown in Fig. 114, by holding the square flat against the side of the stuff with the inside edge of the blade held along the line of the edge of the stuff and the tongue of the square extending squarely across the side of the stuff so that marks made along the inside edge of the tongue of the square (*A–G*, Fig. 114) will be at right angles to the edge of

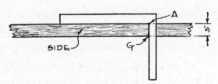

Fig. 114.  Making a Mark Square across Side of
Scantling

the stuff. Having all these marks on the sides and one edge, the stuff can be sawed up into short pieces, each end of which will be beveled so as to fit properly between 2- by 8-inch joists spaced 16 inches on centers. These will be suitable for both pieces in the bridging pair.

Where the floor joists are 2 inches by 10 inches, take a long piece of 2- by 3-inch scantling as before. Find the 9½-inch division mark on the outside edge of the tongue of the steel square and the 14⅜-inch division mark on the outside edge of the blade or body of the square. Using the 9½-inch division on the tongue of the square in the same way as the 7½-inch division mark was used in the foregoing explanation, mark off the piece of scantling and saw it up into short pieces of bridging with both ends beveled to fit between the 2- by 10-inch floor joists.

Where the floor joists are 2 inches by 12 inches, take a long piece of 2- by 3-inch scantling. Find the 11½-inch division on the outside edge of the tongue of the steel square, and the 14⅜-inch division on the outside edge of the blade of the square. Using the 11½-inch division on the outside edge of the tongue of the square just as the 7½-inch and 9½-inch division marks were used in the preceding examples, mark off the piece of scantling as before and saw it up into bridging pieces which in this case will be beveled correctly at the ends so as to fit snugly between the 2- by 12-inch joists.

If the joists were spaced 20 inches apart center to center, instead of 16 inches, it would be necessary to use the 18⅜-inch division (20 inches less 1⅝ inches) on the outside edge of the blade or body of the steel square instead of the 14⅜-inch division; otherwise, the procedure would be the same as described above. If (as is unlikely) the joists were spaced 12 inches apart center to center, the 10⅜-inch division (12 inches less 1⅝ inches) on the outside edge of the blade of the square would be used.

**Spacing Studding.** After the first-floor joists have been set in place, the next work to be done in building a house usually is to lay the rough flooring on top of these joists to form a solid floor or platform on which to work. See Fig. 115 for section through box sill. The rough flooring is laid diagonally and extends to the outside of the box sill (or other sill) all around the outside walls of the house. The vertical studding of the outside walls and the partitions is set above this rough flooring with the lower ends of the vertical studs resting on horizontal 2- by 4-inch scantlings which are laid down flat on the top of the rough flooring. These scantlings are called *sills* or *sole pieces* or *bottom plates*.

In the case of this house being built, imagine that the rough flooring and the sole pieces have been laid and that you are ready to start setting the upright studding in place on top of the sole pieces. Such studding is almost always spaced 16 inches apart, center to center, or from the right-hand side of one stud to the right-hand side of the next stud. This is because wood laths are 4 feet long, and 16 inches is exactly ⅓ of 4 feet (48 inches); so, with the 16-inch spacing of the studs, one end of a 4-foot lath can be placed at the center of a stud and the lath will cross two full studs or three spaces. Thus one end of the lath will be at the center of the third stud from the one supporting the other end of the lath.

To space the upright studding correctly, it is necessary to mark the position of each stud on the top of the sole pieces. This can be done with a rule or a tape but, since it can be done even more easily with the aid of the steel square, the square will be used. Assume that

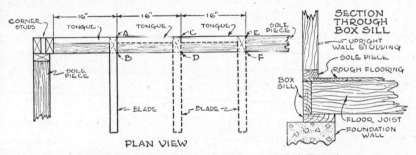

PLAN VIEW

Fig. 115. Spacing Studding with Steel Square (16-Inch Tongue)

the square which you are going to use has a tongue 16 inches long and a blade 2 inches wide; the length of the tongue is therefore the same as the spacing of the studs, and the width of the blade or body is only slightly more than the thickness of a 2- by 4-inch stud. This being the case, and remembering that the thickness of the studs is always parallel with the length of the sole pieces, note in Fig. 115 that if the square is laid down so the 16-inch tongue is flat on the top of the sole piece, with the left-hand end of the tongue tight against the right-hand face of the corner stud, then the 2-inch blade will be at right angles to the length of the sole piece and in such a position that the right-hand edge of the blade (A–B, Fig. 115) will be just 16 inches from the right-hand side of the corner stud.

Now make a mark on the top of the sole piece along the right-hand edge of the blade of the steel square; this mark (A–B, Fig. 115) will locate the position of the right-hand side of the first stud to the right of the corner stud. Move the square along the top of the sole piece to the right until the left-hand end of the 16-inch-long tongue is at the mark which has just been made. The blade of the square is now in such a position that a mark made on the top of the sole piece along the right-hand edge of the blade in its new position (C–D, Fig. 115) will locate the position of the right-hand side of the second stud to the right of the corner stud. By doing the same thing again, you can mark along the right-hand edge of the blade of the square (E–F, Fig. 115) the position of the right-hand side of the third stud from the corner stud. In this way, the positions of any number of studs can be marked out so that the studs will be exactly 16 inches apart center to center.

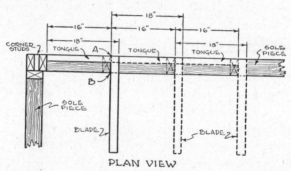

PLAN VIEW

Fig. 116. Spacing Studding with Steel Square (18-Inch Tongue)

If the square has a tongue 18 inches long, the distance from the end of the tongue to that edge of the blade or body which is nearest to the end of the tongue is exactly 16 inches. In using this square for spacing studding, place the square flat on top of the sole piece with the left-hand end of the tongue against the right-hand side of the corner stud, and make a mark on top of the sole piece along the *left-hand* edge of the blade of the square (A–B, Fig. 116). This mark will locate the position of the right-hand side of the first stud to the right of the corner stud.

Now move the square along to the right until the left-hand end of the 18-inch-long tongue is at the mark which has just been made; then make another mark on top of the sole piece along the left-hand edge of the blade of the square. This mark will locate the position of the right-hand side of the second stud to the right of the

corner stud. The position of the third stud to the right of the corner
stud and as many others as necessary can be marked in the same
way when the square has a tongue 18 inches long.

**Preparing Braces for Trussed Openings in Partitions.** Putting
up the studding for the outside walls and the interior partitions of
the house presents no special problems which require the use of the
steel square in their solution, except in the case of unusually wide
openings for doors and windows. Because the walls and partitions
usually support the ends of the floor joists of the floors above them,
it is necessary, wherever there is a wide opening in the stud wall or
partition, to place some diagonal bracing in the space over the head
of the door or window opening to form a sort of truss. This truss
transfers most of the load from above the opening over to the studs
or posts at each side of the opening, where the studs are often
doubled to take the extra burden. The cutting of these diagonal
braces so that they will fit snugly against the upright studding at
each side of the opening and the horizontal or upright members
just over the door or window requires the use of the steel square.

The first-floor plan of the house being built (Fig. 103) shows
that there is an opening 6 feet wide in the partition between the
kitchen and the pantry. This is the sort of opening which should
have diagonal braces forming a truss over it, and they can be cut
to fit with the help of the steel square.

The finished opening is 6 feet wide and 6 feet, 8 inches high.
Allowing 1 inch at each side for the finished frame, the width between studs will be 6 feet, 2 inches, and allowing 1¼ inches at the
head of the opening for the finished frame, the height from the finished floor to the underside of the rough framed header over the
opening will be 6 feet, 9¼ inches. To the top of the rough header
will be 4 inches more, or 7 feet, 1¼ inches. The height of the first
floor from floor to floor is 9 feet, 3 inches; see Fig. 108. Subtract
1¾ inches for rough and finished flooring, 8 inches for second-floor
joists, and 4 inches for the double partition plate; this leaves 8 feet,
1¼ inches as the height from the finished first-floor level to the underside of the double partition plate. Therefore, the height from the top of
the rough header over the door opening to the underside of the partition plate above will be 8 feet, 1¼ inches less 7 feet, 1¼ inches, which
is just 1 foot. This 1 foot is the *rise* of the brace over the opening.

The clear width of the space between studs over the door opening is 6 feet, 2 inches. If an upright 2- by 4-inch stud is placed over the opening in the center of this space, as shown in Fig. 117, it will divide the space into two equal parts, each 3 feet wide (6 feet, 2 inches less 2 inches divided by 2); thus, the *run* of each of the two braces will be 3 feet. If the lower ends of the braces bear against the top of the lintel or header over the door opening, and the upper ends of the braces bear against the sides of the upright stud over the center of the door opening, as shown in Fig. 117, then the top side of the brace will run diagonally across the half opening from the lower outside corner to the upper inside corner and the length

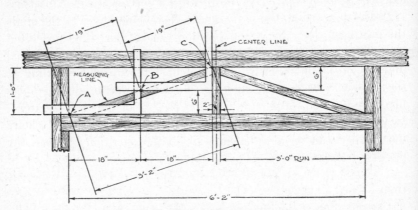

Fig. 117. Bracing over Wide Opening for Door

required for the brace will be the length of this side of the brace. The line of intersection of this side of the brace with the edge of the brace (*A–C*, Fig. 117) can be used as a measuring line. One-half of the *run* of the brace will be one-half of 3 feet, or 18 inches, and one-half of the *rise* of the brace will be one-half of 1 foot, or 6 inches.

As these braces are not very long, they can be cut from a piece of 2- by 4-inch scantling.

Note in Fig. 117 that if the square is applied to the edge of the brace as shown in Fig. 117, with the two division marks on the measuring line 18 inches from the heel of the square on the outside edge of the blade or body and 6 inches from the heel on the outside edge of the tongue of the square, the blade of the square will give the cut for the lower end of the brace to fit it against the top side of the header over the door opening, as at point *A*, Fig. 117. Fig. 117 also shows that the point is marked on the measuring line of the brace where the 6-inch division on the outside edge of the tongue of the square crosses it (point *B* in Fig. 117). Then the square is moved along the edge of the brace to the right until the 18-inch division on the outside edge of the blade or body is at the mark on the measuring line where the outside edge of the tongue crossed it before (point *B*, Fig. 117). The outside edge of the tongue in its new position will give the cut for the upper end of the brace to fit it against the upright stud over the center of the door opening, provided that the 6-inch division mark on the outside edge of the tongue of the square has been kept on the measuring line, as at point *C*, Fig. 117. Fig. 117 also shows that the true length of the brace measured along the measuring line (the upper edge) will be just twice the distance measured across diagonally on the steel square between the 18-inch division on the outside edge of the blade and the 6-inch division on the outside edge of the tongue of the square. This distance, measured exactly with a rule or with the edge of another square, is found to be very nearly 19 inches—so near that it can be taken as 19 inches. Then the length of the brace is twice 19 inches or 38 inches, which is 3 feet, 2 inches.

To cut the brace, therefore, take a piece of 2 by 4 a little longer than 3 feet, 2 inches—say 3 feet, 6 inches. Place the steel square flat against one of the 2-inch edges of the stuff as shown in Fig. 118, in such a way that the 18-inch division on the outside edge of the blade will rest 2 inches from the left-hand end of the piece of stuff and on the edge of the stuff farthest from the heel of the square, while the 6-inch division on the outside edge of the tongue of the square also rests on the same edge of the stuff (that is, on the measuring line). Make a mark (*B–C*, Fig. 118) across the 2-inch edge of the stuff along the outside edge of the blade or body of the square. This mark will show the cut to be made for the lower end of the brace. Also mark the place (point *D*, Fig. 118) where the 6-inch division on the outside edge of the tongue of the square rests on the edge of the stuff farthest from the heel of the square (the measuring line). Move the square to the right along the edge of the stuff, taking care to keep the 18-inch division of the outside edge of the blade and the 6-inch division of the outside edge of the tongue always on the measuring line. This operation can be made easier by using a *fence* on the square, as explained at the beginning of Chapter 2. Having moved the square along the edge of the stuff as just described, make a mark across the 2-inch edge of the stuff along the outside edge of the tongue of the square (*E–F*, Fig. 118). This mark will show the cut to be made for the upper end of the brace.

**Common Rafters.** After the studding for the walls and partitions has been set up and the floor joists and rough flooring are in place, the next thing is to erect the roof framing. Before this can be done, the roof rafters have to be cut to fit properly into place in the roof frame. In the house being built, there are different kinds of rafters, all of which must be cut to the right length with the ends fitted to rest on the wall plate at the bottom and against the ridge

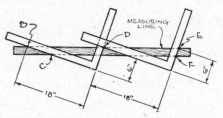

Fig. 118.  Laying Out Brace with Steel Square

or other rafters at the top. Without the steel square, this job would be much harder than it is when the square is employed.

The first rafters to be considered are those forming the roof over the maid's room, etc., at the east end of the house. See Figs. 103 and 108. This is a simple gable end roof with only common rafters. The span out-to-out of wall plates is 16 feet, 4½ inches so that the *run* of the rafters is one-half of this or 8 feet, 2¼ inches. The *rise* of the roof is 10 feet, 9½ inches from the eave line (in which the plane of the outside of the wall studs would cut the slope of the roof) up to the top of the ridge. This is slightly steeper than a five-eighths pitch, and the rise per foot run will be 10 feet, 9½ inches (or 129.5 inches) divided by 8 feet, 2¼ inches (or 8.1875 feet), which is about 15¹³⁄₁₆ inches. Since this rise per foot run is not an even number of inches, the length per foot run of the rafter cannot be taken directly from the rafter tables on the steel square as explained in Chapter 3.

**Determining Length, Using Steel Square.** There is, however, another way to get the length of the rafter by use of the steel square as follows: The run of the rafter divided by 12 is $8\frac{3}{16}$ inches, and the rise divided by 12 is nearly $10\frac{13}{16}$ inches. Find the $8\frac{3}{16}$-inch division on the outside edge of the tongue of the square and the $10\frac{13}{16}$-inch division on the outside edge of the blade of the square. With a rule, measure off the diagonal distance from one of these divisions across to the other as accurately as you can. This distance is found to be a little more than $13\frac{1}{2}$ inches; see Fig. 119A. Since one-twelfth of the rise of the rafter and one-twelfth of the run were taken, the result (namely, $13\frac{1}{2}$ inches) must be one-twelfth of the entire length of the rafter. Therefore, multiply $13\frac{1}{2}$ inches by 12 and obtain 13 feet, 6 inches. The length of the rafter along the measuring line, or along the *back* of the rafter, is a little more, say 13 feet, $6\frac{1}{2}$ inches.

**Allowance for Tail and Seat.** The rafter will actually be a foot or two longer than this from end to end because of the tail at the eaves and extra length at the ridge. Therefore, select a piece of 2- by 6-inch stuff about 16 feet long and smooth up one edge of it. The dressed edge will be the top edge or *back* of the rafter. Using a gauge, mark or scratch off a short measuring line near one end on one side of it, $3\frac{1}{2}$ inches from one edge, after having dressed the edge of the stuff. A point on this measuring line 1 foot, 3 inches from the left-hand end of the piece of stuff will allow for the *tail*, as shown in Fig. 119A. This mark (*A*, Fig. 119A) will correspond to the upper outside corner of the wall plate. It must be far enough away from the edge of the stuff which will form the lower edge of the rafter so that, when the seat cut is made, the rafter will have a long enough *seat* or bearing on the top surface of the wall plate (*A–C*, Fig. 119A).

To test this, lay the steel square down flat across the side of the stuff with the dressed edge of the stuff away from you, the heel of the square pointing toward you, and the blade of the square at your right hand. Place the $8\frac{3}{16}$-inch division of the outside edge of the tongue at point *A*, Fig. 119A, and the $10\frac{13}{16}$-inch division of the outside edge of the blade on the measuring line farther along to the right, as shown in Fig. 119A. The distance measured along the outside edge of the tongue of the square toward the heel from point $A$ to the edge of the stuff at point $C$, will be the length of the seat. The width of the seat will be the thickness of the rafter. While the square is in this position, a mark will be made along the outside edge of the tongue to locate the seat cut. The length of the bearing on the top of the wall plate ($A$–$C$, Fig. 119A) will be a little more than 3 inches, which is enough.

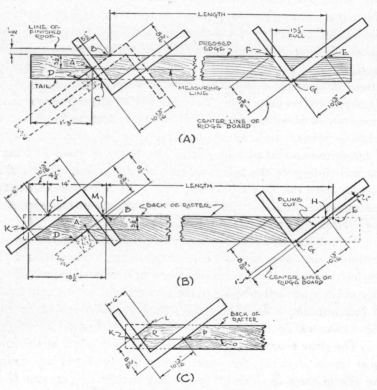

Fig. 119. Laying Out Common Rafters with Steel Square

**Marking for Cuts.** Now place the square on the stuff with the heel pointing away from you and the blade at your left, with the $10\frac{13}{16}$-inch division of the outside edge of the blade on point $A$ as shown by dotted lines in Fig. 119A, and with the $8\frac{3}{16}$-inch division of the outside edge of the tongue of the square on the measuring line farther along to the right. Make a mark ($B$–$A$–$D$, Fig. 119A) across the side of the stuff along the outside edge of the blade of the square when it is in this position. The portion of this line between point $A$, Fig. 119A, and that edge of the stuff which will be the lower edge of the rafter at point $D$ will locate the vertical cut at the seat of the rafter where it fits against the outside face of the wall plate.

The portion of this line between point $A$ and that edge of the stuff which will be the top edge or back of the rafter is the line in which the outside face of the wall studs would cut the rafter, if this outside surface of the wall studs were extended up above the top of the wall plate; and point $B$, Fig. 119A, locates where the outside face of the wall studding would cross the top surface of the rafters, if this face were extended.

As stated on page 36 in this book, the top or back of the rafter also can be used as a measuring line. Starting at point $B$, Fig. 119A, measure off to the right along the dressed edge of the stuff the length of the rafter, and thus locate point $E$, Fig. 119A (the point in which the top surface of the rafter would cross the center of the ridge board). This can be checked by using a fence on the steel square and stepping the square along the entire length of the stuff as illustrated in Chapter 3.

Now place the square on the stuff with the $10^{13}\!/_{16}$-inch division of the outside edge of the blade on point $E$, Fig. 119A, and the $8^{3}\!/_{16}$-inch division of the outside edge of the tongue on the top edge of the stuff farther along to the left at point $F$. With the square in this position, a line along the outside edge of the blade $E$–$G$, Fig. 119A, will be the center line of the ridge board.

The ridge board is 2 inches thick, and the top cut of the rafter must be made so that the upper end of the rafter will rest against the side of the ridge board, 1 inch away from the center line of the ridge. For this reason it is necessary to move the square back to the left, as shown in Fig. 119B at $H$, until an allowance of 1 inch for one-half the thickness of the ridge board has been made. When the square is in this position, make a mark across the side of the stuff along the outside edge of the blade of the square ($H$–$G$, Fig. 119B). This mark will locate the top cut or *plumb* cut for the upper end of the rafter. After this top cut has been made, the rafter will be complete except for the shaping of the tail to conform to the design shown in Fig. 108.

**Shaping the Tail.** To shape the tail of the rafter, note, by reference to Fig. 108, that the distance measured along the top edge or back of the rafter from point $B$ to the end of the tail of the rafter is 14 inches. Therefore, as shown in Fig. 119B, measure off with a rule 14 inches from point $B$ to the left along that edge of the stuff which will be the back of the rafter, and mark point $L$, which represents the end of the tail on this edge of the rafter. Now hold the steel square in such a way that the heel is pointing away from you with the blade at your left, and place it on the side of the stuff with the $10^{13}\!\!/_{16}$-inch division of the outside edge of the blade at point $L$ (Fig. 119B) and the $8^{3}\!\!/_{16}$-inch division of the outside edge of the tongue on that edge of the stuff which will be the back of the rafter farther along to the right at the point $M$. Make a mark across the side of the stuff along the outside edge of the blade of the square from point $L$ toward the opposite edge of the stuff ($L$–$K$). This mark will show the vertical (plumb) cut for the end of the tail. Before lifting the square, note that the distance from the heel of the square to the line $B$–$A$–$D$, Fig. 119B, is $8\frac{1}{2}$ inches, which is the distance from the outside face of the wall plate to the plumb cut at the end of the tail, measured level. For some rafters, the cut at the end of the tail would be made along the line $L$–$K$, and this would finish the work. However, Fig. 108 shows that the end of the tail for this rafter is shaped differently—that is, the cut along $L$–$K$ in Fig. 119B is only 6 inches long, and the end of the tail is finished with a horizontal cut as well as the vertical (plumb) cut. Therefore, measure with a rule 6 inches along $L$–$K$, Fig. 119B, and thus find point $K$, where the vertical cut and the horizontal cut at the end of the tail of the rafter meet.

Now take a gauge and set it so that you can mark on the side of the stuff a line passing through point $K$ and parallel to that edge of the stuff which will be the top edge or back of the rafter. With the gauge, mark the line $K$–$O$, Fig. 119C. Having located point $K$ and line $K$–$O$, hold the steel square so that the heel is pointing toward you and place it on the side of the stuff in such a position that the $8^{3}\!\!/_{16}$-inch division of the outside edge of the tongue of the square will come at point $K$, and the $10^{13}\!\!/_{16}$-inch division of the outside edge of the blade will rest on the line $K$–$O$ farther along to the right, as shown at point $P$, Fig. 119C. With the square in this position, a line drawn along the outside edge of the tongue from point K to that edge of the stuff which will be the lower edge of the rafter will mark the horizontal cut at the end of the tail of the rafter as shown by line K-R, Fig. 119C.

One rafter is now completely cut, and, using this as a pattern, all of the common rafters for this section of the roof can be cut. In this way it will not be necessary to measure off the length for them or dress the edge for use as a measuring line.

For the gable end roof over the wing at the west end of the house and the two gable end roofs on the north side over the front entrance, the *slope* (rise per foot run) is the same as it is for the wing at the east end; but the spans and therefore the runs of the common rafters for these roofs are in each case different. (The rise per foot run of the common rafters over the main portion of the house also is 15¹⁹⁄₁₆ inches.)

**Hip Rafters.** The next task in connection with which the steel square will be a help is the cutting and shaping of the hip rafters. There are four of these in the frame of the main roof. This is a roof of *even pitch* which means that the two sides and two ends of the roof are of the same slope so that, in the roof plan (Fig. 105), the lines which represent the hips, E-F, F-G, H-I, H-K, make angles of 45 degrees with the lines which represent the eaves or the outside faces of the wall plates, lines M-E, E-I, I-L, K-N, N-G.

**Determining Run and Length of Hip Rafters.** As these hip rafters are quite long, they will be made from 4- by 10-inch stuff, and four pieces of this size will be chosen from which to cut the hip rafters. At this point, the problem of deciding how long a piece of stuff to choose arises, and the square can be used in working it out. However, to arrive at the length, it will first be necessary to determine the run. Because the roof is a roof of even pitch, there will be a *run* of 17 inches for the hip rafter for every foot of run of the common rafter. The run of the common rafter on the main part of the house is about 14 feet and 1 inch (half the distance out-to-out of wall plates, which means stucco is not counted). Seventeen inches is 1⁵⁄₁₂ feet, and 14¹⁄₁₂ times this is about 20 feet, so 20 feet will be the run of the hip rafters.

This can be checked with the steel square by finding the 14¹⁄₁₂-inch division on the outside edge of the back of the tongue of the

square, and the $14^1/_{12}$-inch division on the outside edge of the blade
of the square, and measuring the diagonal distance across between
these two points (as shown in Fig. 120) by means of a rule or the
edge of another steel square. This distance will be found to be about
20 inches. In doing this, one inch is taken on the square for each
foot in the run of the rafters.

The rise (see Fig. 110) of the roof on the main part of the building, and of the common and hip rafters as well, is 18 feet, 6 inches.
Therefore, to find what the *length* of the hip rafter will be (very
nearly), take the 18-inch division at the end of the outside edge of
the tongue of a square having a tongue 18 inches long (adding 6

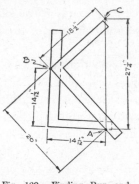

Fig. 120. Finding Run and
Length of Hip Rafter with
Steel Square

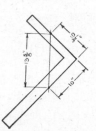

Fig. 121. Dimensions
Divided by Two When
Square Has 16-Inch
Tongue

inches to represent the 18-foot, 6-inch rise of the hip rafter) and the
20-inch division on the outside edge of the blade of the square (to
represent the 20-foot run of the hip rafter) and measure with a rule
the diagonal distance across between these two points. See Fig. 120.
The distance will be about $27\frac{1}{4}$ inches and so the length of the hip
rafter will be about 27 feet, 3 inches.

If the square which you are using has a tongue only 16 inches
long, take half of each of these dimensions; that is, $9\frac{1}{4}$ inches on
the tongue and 10 inches on the blade as shown in Fig. 121. Measure
off the distance between them, $13\frac{5}{8}$ inches, and double it to get
$27\frac{1}{4}$ inches.

To this length of $27\frac{1}{4}$ feet must be added about 4 feet to allow
for the tail of the hip rafter, as well as the extra length at the upper

end, making 31 feet as the approximate total length for the hip
rafter. Therefore, obtain a few pieces of 4- by 10-inch stuff, 31 feet
long, and cut and shape one of these and use it as a pattern for the
other hip rafters.

**Marking for Seat Cut of Hip Rafter.** First, the seat cut can
be made so the hip rafter will rest on top of the wall plates which
meet at the corner of the house and will also fit over the wall plates
to form the tail of the rafter. It is necessary before making a cut
to consider the length of the tail so the seat cut will be far enough
from the end of the stuff to leave plenty of length projecting beyond
the seat cut to form the tail. Referring back to the common rafter,
section view on Fig. 108, note that the end of the overhanging
*common* rafter at the eaves is $8\frac{1}{2}$ inches from the outside face of the
wall plate, measured level or horizontally. The end of the overhang-

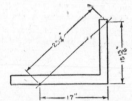

Fig. 122. Finding Length of
Hip per Foot Run of Common
Rafter

ing *hip* rafter will project more than $8\frac{1}{2}$ inches from the outside
face of the wall plate, in fact $^{17}/_{12}$ times more, but to make sure of
having enough take $^{17}/_{12}$ times 12 inches or 17 inches. Propor-
tionately, then, 17 inches should provide plenty of horizontal distance
between the end of the *hip* rafter and the outside face of the wall
plate. Now you can determine what length to allow for the tail of
the hip rafter. Since your 12-inch common rafter run becomes 17
inches on a hip rafter, and you are taking 17 inches as the horizontal
overhang of the hip rafter, the corresponding rise of the hip rafter
will be $15\frac{13}{16}$ inches. Find the 17-inch division on the outside edge
of the blade of the steel square and the $15\frac{13}{16}$-inch division on the
outside edge of the tongue of the square (representing the rise per
foot run of the common rafter of the main part of the house), and
measure with a rule the diagonal distance between these two points
on the square. See Fig. 122. The distance will be found to be about

23¼ inches (which represents the length of the hip rafter for every foot of the common rafter). To this, add half the depth of the rafter, or 5 inches, making 28¼ inches.

As a result of the foregoing calculation, make a mark (*B*, Fig. 123) on that edge of the piece of stuff which will be the top edge or back of the hip rafter at least 32 inches from the left-hand end of the stuff when the stuff is lying on its side with the marked edge away from you. Now hold the steel square so that the heel is pointing toward you, with the tongue in the right hand and the blade in the left hand. Place the square down flat on the side of the stuff so that the 15¹³⁄₁₆-inch division on the outside edge of the tongue will rest at the mark *B* which has just been made on that edge of

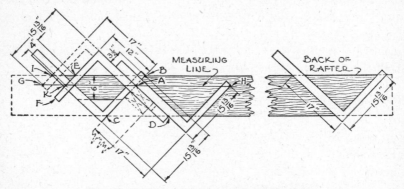

Fig. 123. Steel Square Used for Laying Out Hip Rafter

the stuff which will be the top edge or back of the hip rafter, while the 17-inch division on the outside edge of the blade of the square will rest on the same edge of the stuff farther along to the left, as shown at *I* in Fig. 123. With the square in this position, make a mark across the side of the stuff along the outside edge of the tongue of the square. This line, *B–A–C* in Fig. 123, shows where the center of the hip rafter will be set over the vertical meeting line of the two outside faces of the wall plates which meet at the corner of the building.

Somewhere on this line, a point represents the position of the outside upper corner or edge of the wall plate, which will also be the outer end of the seat cut for the hip rafter. This point will have to be located to make the seat cut; see point *A*, Fig. 123. First, however, it must be decided whether or not to back the hip rafter, and you decide not to back it. Because of this, the horizontal cut of the seat cut (which makes the lower end of the hip rafter fit on top of the wall plate) will have to be made a little nearer to the top edge of the hip rafter than if the rafter were going to be backed, so that the top edge of the rafter will not stick up above the under side of the roof boarding and the top edges of the jack rafters, as shown in Fig. 63. This means that in Fig. 123 the distance $A-B$ (from the upper outside corner or edge of the wall plate straight up to the top edge of the hip rafter), instead of being the same as the distance $A-B$ in Fig. 119A for the common rafters (which would be similar to those of the maid's room), will be a little less than this. How much less depends upon the rise per foot run of the common rafters and the thickness of the hip rafter.

In this roof the rise per foot run of the common rafter is $15^{13}\!/_{16}$ inches and the thickness of the hip rafter is a little less than 4 inches. Half the thickness of the hip rafter is a little less than 2 inches. The rule is to take half the thickness of the hip rafter, multiply it by the rise per foot run of the common rafter, and divide the result by 17. This is the same as saying, divide half the thickness of the hip rafter by 17 and multiply the result by the rise per foot run of the common rafter. The calculation can be worked out by arithmetic, but the same result can be reached by using the steel square.

To do it, take the square and lay it flat down with the face uppermost on a large piece of paper on the floor; see Fig. 124. Draw lines on the paper along the outside edges of the blade and the tongue of the square meeting at the heel of the square. Find the 17-inch division on the outside edge of the blade and make a mark there ($B$, Fig. 124). Find the $15^{13}\!/_{16}$-inch division (equal to the rise per foot run of the common rafter) on the outside edge of the tongue of the square and make a mark there ($C$, Fig. 124). Lift the square and join the two marks with a line $B-C$ in Fig. 124. Extend the line $A-B$ farther to the right. Then lay the square down on the paper again with the outside edge of the blade on the line $A-B$ and move the square over to the right far enough so that the 2-inch division on the outside edge of the blade (equal to one-half the thickness of the hip rafter) will come at the point $B$, as shown by the dotted lines outlining the square in Fig. 124. The outside edge of the tongue of the square in its new position (shown dotted in *E–F*, Fig. 124) will now be parallel to the line *A–C*, which was drawn along the outside edge of the tongue of the square in its first position (shown by full lines in Fig. 124); and this outside edge of the tongue in its new position (*E–F*, Fig. 124) will cross the line *B–C* at point *D*, which should be marked. Note in Fig. 124 that the distance *D–E* (from the mark *D* to the heel of the dotted square) is very nearly 2 inches, and this is the distance you are trying to find. This is the amount by which distance *A–B* in Fig. 123 is less than distance *A–B* in Fig. 119A. Distance *A–B* in Fig. 119A is 5½ inches; therefore, in Fig. 123, point *A* is 3½ inches from point *B* on the line *B–C* as shown in Fig. 123 and indicates where the side of the hip rafter would cross the outside upper edge of the wall plate at

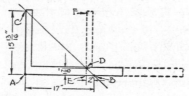

Fig. 124.   Finding Amount to Drop Hip
Rafter, Using Steel Square

the corner of the building. By means of a gauge, mark on the side of the hip rafter a line (*A–H*, Fig. 123) passing through point *A* and parallel to the top edge of the rafter. This line will be the measuring line.

   To finish making the seat cut, hold the steel square so that the heel is pointing toward you with the blade at the left, and place the square on the side of the stuff with the 17-inch division of the outside edge of the blade at point *A* as shown in Fig. 123, with the 15¹³⁄₁₆-inch division of the outside edge of the tongue of the square on the measuring line farther along to the right at point *H*. With the square in this position, a mark made across the side of the stuff along the outside edge of the blade of the square, *A–D* in Fig. 123, from point *A* to the edge of the stuff nearest to the heel of the square will show the cut to be made to fit the hip rafter over the top face of the wall plate. The other part of the seat cut as previously determined will be made along the line *A–C*, Fig. 123, from point *A* to that edge of the stuff which will be the lower edge of the hip rafter.

**Marking Cuts for Tail of Hip Rafter.** After the seat cut has been made, mark off the cuts for the tail of the hip rafter in order to finish the lower end of this rafter. As this hip rafter is fairly deep, it can be done by applying the square to the side of the stuff and marking off the cuts on it. A measuring line has already been marked off on the side of the stuff (line $A$–$H$, Fig. 123) parallel to the top edge or back of the rafter, and passing through point $A$, which is the point where the hip rafter would intersect the outside upper edge of the wall plate. Also the line $A$–$C$ has been established (Fig. 123) which is the line of intersection between the side of the hip rafter and the outside face of the wall studding and the outside face of the wall plate.

Now hold the square with the heel pointing away from you and with the blade to your right and place it on the side of the stuff so that the 17-inch division of the outside edge of the steel square blade is at point $A$, Fig. 123, while the $15^{13}/_{16}$-inch division of the outside edge of the tongue of the square (equal to the rise per foot run of the common rafters) is on the measuring line farther along to the left, at point $G$. With the square in this position, shown by dotted lines in Fig. 123, the outside edge of the blade of the square is at right angles to the line $A$–$C$ representing the outside face of the wall, and would be level if the stuff were raised up into the sloping position which the hip rafter will occupy in the actual roof frame.

Refer back to the work done on the common rafter shown in Fig. 119B and remember that the distance, measured level from the outside face of the wall plate to the plumb cut at the end of the tail of the rafter, was $8\frac{1}{2}$ inches. The similar distance for the tail of the hip rafter would be measured along the outside edge of the blade of the square when the square is in the position shown dotted in Fig. 123. It would be more than $8\frac{1}{2}$ inches because the hip rafter is at an angle of 45 degrees with the wall plate instead of at right angles to the wall plate as the common rafter was. To determine what this distance is, find the $8\frac{1}{2}$-inch division of the outside edge of the blade of the steel square and the same division of the outside edge of the tongue of the square and with a rule measure diagonally

across between these two points on the square. The distance will
be found to be 12 inches, as shown in Fig. 125. To make the end
of the tail of the hip rafter line up with the ends of the tails of
the common rafters, this distance for the hip rafter will be used in
combination with the steel square as follows:

Holding the square so that the heel is pointing away from you,
with the blade at your right, place the square on the side of the stuff
as shown in Fig. 123 in such a way that the 12-inch division of the
outside edge of the blade of the square (equal to the distance just
found) is on point A, Fig. 123, while the outside edge of the blade
is on the line A–D drawn across the side of the stuff. When the square
is in this position, a line drawn across the side of the stuff along the
outside edge of the tongue of the square (line E–F, Fig. 123) will

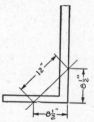

Fig. 125. Square Used to Find
Overhang for Tail of Hip Rafter

show the position for the plumb cut for the end of the tail of the hip
rafter. The end of the stuff will be cut off along this line.

For many rafters, this would complete the work on the tail
but, for this particular hip rafter, there is another cut to be made in
addition to the plumb cut, so that the lower edge of the tail of the
hip rafter will be as horizontal or level as the tails of the common
rafters shown in Fig. 119C. In Fig. 119C, notice that the depth
of the plumb cut at the end of the tail of the common rafter from
point L to point K is 6 inches measured straight down from the top
edge or back of the common rafter. For the hip rafter, this depth
of the plumb cut at the end of the tail of the rafter would be the
same as for the common rafters except that, as you recall, when the
seat cut for the hip rafter was made, the stuff was cut so that the top
edge or back of it would be dropped straight down a distance of 2
inches to avoid backing the hip rafter. Because of this, the depth
of the plumb cut at the end of the tail of the hip rafter, measured

from that edge of the stuff which will become the top edge or back
of the rafter (distance $E-K$ in Fig. 123), will then be 2 inches less
than the similar distance for the common rafter (distance $L-K$ in
Fig. 119B).

Therefore, having found the line for the plumb cut for the tail
of the hip rafter (line $E-F$ in Fig. 123) and point $E$ on that edge of the
stuff which will be the top edge or back of the hip rafter, measure
off with a rule from point $E$, Fig. 123, the distance $E-K$, equal to
6 inches less 2 inches, or 4 inches, showing the depth of the plumb
cut for the tail of the hip rafter and giving us the position of point
$K$ from which the horizontal or level part of the end cut can be marked
off for the tail of the hip rafter.

To make this cut, do just as you did for the similar cut at the

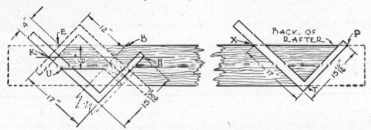

Fig. 126. Laying Out Tail and Plumb Cut for Hip Rafter with Steel Square

end of the tail of the common rafter (Fig. 119C) except use different
dimensions. Take a gauge and set it so it will mark on the side of
the stuff a line which will pass through point $K$ and be parallel to
that edge which will be the top edge or back of the hip rafter. Mark
off a line about 2 feet, 6 inches long ($K-H$ in Fig. 126). Then hold
the square so that the heel is pointing toward you with the blade
to the left, and place it on the side of the stuff so that the 17-inch
division of the outside edge of the blade of the square will come at
point $K$, Fig. 126, and the $15\frac{13}{16}$-inch division of the outside edge of
the tongue of the square (equal to the rise per foot run of the common rafters) will rest on the line $K-H$ farther along to the right, as
shown at point $H$. With the square in this position, draw a line
along the outside edge of the blade from point $K$, Fig. 126, to that
edge of the stuff which will be the lower edge of the hip rafter. The
overhanging portion of the hip rafter will be cut down to a width
of 6 inches, as shown in Fig. 126. A cut made along this line ($K-U$,

Fig. 126) will be the horizontal cut for the end of the tail of the hip rafter so it will match the ends of the common rafters. This completes the cuts to be made for the lower end of the hip rafter, but the cuts for the upper end of the rafter where it rests against the ridge board still have to be marked out and made.

**Marking Cuts for Upper End of Hip Rafter.** Before marking out the cuts at the upper end of the hip rafter, it will be necessary to find a working point from which to start. To do this, work on the edge of the stuff which will be the top edge or back of the hip rafter. This edge, shown in Fig. 127, has been dressed. The mark at point $B$, Fig. 127, has already been made on one edge of the back of the rafter, where the back and side of the rafter meet. Place the square flat against the back of the rafter, with the inside edge of the blade against the side of the stuff and the tongue of the square extending straight across the back of the rafter at point $B$, Fig.

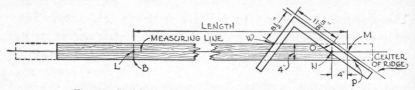

Fig. 127. Steel Square Used to Lay Out Side Cut for Hip Rafter

127; make a mark from $B$ straight across the edge of the stuff along the inside edge of the tongue of the square. This mark will be at right angles to, or square with, the edges of the back of the rafter. Mark on this line by measurement with a rule the exact center of the edge of the stuff at point $L$. With a gauge, mark off a line ($L$–$N$–$M$, Fig. 127) for the full length of the stuff along the center of its edge passing through the point $L$ and parallel to the two edges of the back of the rafter. This line will mark the center line of the back of the hip rafter and it will be your measuring line. The length of the rafter will be measured off along this line.

Because the pitch of this roof is such that the rise per foot run is not an even number of inches ($15\frac{13}{16}$ inches), it is not possible to make use of the rafter tables on the steel square to find the theoretical length of the hip rafter, that is, the distance measured along the measuring line from point $L$, Fig. 127, which is plumb above the outside upper corner of the wall plate, up to point $M$, where the measuring line on the back of the hip rafter (line $L$–$N$–$M$ in Fig. 127) would meet the center of the ridge board. However, this length can be found by using the steel square in another way.

It has already been determined that the run of the hip rafter is 20 feet and that its total rise is $18\frac{1}{2}$ feet. Repeating the procedure used in Fig. 120, find the 20-inch division of the outside edge of the blade or body of the square and mark it on a line drawn along the outside edge of the blade of the square while it is laid flat on a large piece of paper (point $A$, Fig. 120). Also draw a line ($B$–$C$, Fig. 120) on the paper along the outside edge of the tongue of the square before the square is lifted from the paper. By extending this line a few inches, you will be able to mark on it the place where the $18\frac{1}{2}$-inch division of the outside edge of the tongue of the square would come if the tongue of the steel square were 19 inches long instead of 16 inches (point $C$, Fig. 120). After having marked these two points, take a rule and measure the diagonal distance across between them. This distance ($A$–$C$, Fig. 120) will be very nearly $27\frac{1}{4}$ inches. Since an inch was taken on the edge of the square for each foot in the run and rise of the hip rafter, it is possible to say that the theoretical length of the hip rafter is 27 feet, 3 inches. (To check this length, remember that the length of the hip rafter for every foot of the common rafter run was found to be $23\frac{1}{4}$ inches—see Fig. 122—and that the run of the common rafter was $14^1/_{12}$ feet. Therefore, multiplying $23\frac{1}{4}$ inches by $14^1/_{12}$, or 14.08, will give a figure which can be used as the rafter length. It is 327.36 inches or nearly 27 feet, 3 inches. This is the length of the hip rafter.) Point $L$, Fig. 127, has already been determined, so measure off along the measuring line on the back of the hip rafter a distance of 27 feet, 3 inches. In this way the position of point $M$, Fig. 127, will be found on the measuring line at the upper end of the back of the hip rafter.

To allow for the thickness of the ridge board (which is of 2-inch stuff) and the thickness of the hip rafter itself, measure back with the square along the measuring line on the back of the hip rafter from point $M$, Fig. 127, a distance of 4 inches, the thickness of the hip rafter, and get point $N$ on the measuring line. With the inside edge of the blade against the side of the stuff and the tongue extending square across the back of the rafter with its inside edge on point $N$, make a mark ($N$–$O$, Fig. 127) along the inside edge of the tongue of the square straight across the back of the hip rafter passing through point $N$ at right angles to the measuring line. This mark will give you the point $O$ on that edge of the back of the hip rafter which will be nearest to the end of the ridge. It is through this point that the side cut and plumb cut for the upper end of the rafter will be made.

To find a similar point on the other edge of the back of the hip rafter, make use of the steel square as follows: Hold the square so that the heel of the square is pointing away from you, with the blade at your right. Then place the square flat against the edge of the stuff which will be the top edge or back of the hip rafter, in such a position that the $11\frac{5}{8}$-inch division of the outside edge of the blade of the square will come at point $O$, Fig. 127, and the $8\frac{1}{2}$-inch division of the outside edge of the tongue of the square will come on the same edge of the back of the rafter, farther along to the left at point $W$. Notice that $11\frac{5}{8}$ inches is just one-half of the length of the hip rafter per foot run of the common rafter ($23\frac{1}{4}$ inches) and that the $8\frac{1}{2}$ inches is just one-half of the run of the hip rafter per foot run of the common rafter (17 inches). Each of these dimensions is divided by two because the blade and tongue of the square are not long enough to allow the use of the full dimensions. This is the rule whenever the rise per foot run of the roof is such that the rafter tables on the steel square cannot be used.

Now, with the square in this position on the edge of the stuff, mark a line ($O$–$P$, Fig. 127) across the edge of the stuff along the outside edge of the blade of the square from point $O$ to the other edge of the back of the rafter which this mark will cross at point $P$, Fig. 127. This mark shows where the back of the hip rafter will have to be cut to fit it against the ridge board. It will not do, however, to cut the stuff through on this line square with the top edge. In order to mark out the plumb cuts which must be made, again use the square.

To mark off the plumb cut on that side of the stuff which will be the side of the hip rafter nearest to the hipped end of the roof, take the stuff and lay it down on a pair of horses as shown in Fig. 35, with that edge toward you which will be the top edge or back of the hip rafter; see Fig. 128. The point $O$, Fig. 128, will then be on the edge of the stuff nearest to you. Hold the square so that the heel is pointing away from you with the blade to the left. Then place the square flat down on the side of the stuff with the 15¹³⁄₁₆-inch division of the tongue of the square (equal to the rise per foot run of the common rafters) on the mark $O$, Fig. 128, at the edge of the stuff, and the 17-inch division of the outside edge of the blade on the same edge of the stuff (the edge nearest to you) farther along to the left at point $R$, Fig. 128. With the square in this position, make a mark across the side of the stuff along the outside edge of the tongue of the square from point $O$ to the other edge of the stuff (which will be the lower edge of the hip rafter when it is in its position in the roof frame). This mark will be along the line $O–S$ in Fig. 128 and will show where the plumb cut should be made on this side of the stuff to fit the hip rafter against the ridge board.

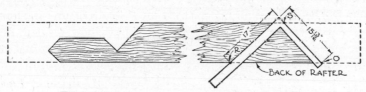

Fig. 128. Steel Square Used for Plumb Cut for Hip Rafter

Now the side cut and plumb cut have been marked out for the upper end of the hip rafter on the edge and one side of the stuff. To mark out the plumb cut on the other side of the stuff (that side of the hip rafter which is farthest away from the hipped end of the roof), use the steel square as follows: Turn the stuff over on the sawhorses so that the edge which will be the top edge or back of the hip rafter will be farthest away from you, as shown at the right-hand end of Fig. 126. Point $P$, Fig. 127, where the side cut on the back of the rafter would meet this side of the rafter, has already been marked; this shows as point $P$, Fig. 126.

Hold the steel square so that the heel will be pointing toward you, with the blade at your left, and place the square flat down on the side of the stuff with the 15¹³⁄₁₆-inch division of the outside edge of the tongue of the square on point $P$, Fig. 126, and the 17-inch division of the outside edge of the blade on the same edge of the stuff (that edge which will be the top edge or back of the hip rafter) farther along to the left, at point $X$. With the square in this position,

make a mark across the side of the stuff along the outside edge of
the tongue of the square from point P, Fig. 126, to the other edge
of the stuff (which will be the lower edge of the hip rafter when it
is in its position in the roof frame). This mark P–T, Fig. 126, will
show the plumb cut to be made for the side of the hip rafter which
will be farthest from the hipped end of the roof.

Now the side cut and the two plumb cuts have been marked
out for the upper end of the hip rafter as shown at C–D and D–E
in Fig. 57, and cutting the stuff along these marks will complete one
hip rafter. This rafter may be used as a pattern for cutting the other
three hip rafters, one of which will be exactly the same as the pattern
and the other two *opposite hand* to the pattern.

To cut the opposite hand rafter, place the stuff on edge on the
sawhorses close beside the pattern rafter, so that when cut the two
rafters will be as shown in Fig. 129. Point P, Fig. 129, and the plumb

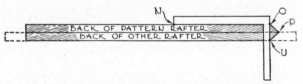

Fig. 129.  Use of Steel Square and Pattern for Opposite Hand Rafters

cut on that side of the rafter which is close against the pattern, can
be easily marked off from the pattern. To get point U, Fig. 129,
on that edge of the stuff which will be the top edge or back of the
rafter, take the steel square and lay it down flat on the edges of the
pattern and the other rafter, with the inside edge of the blade of
the square, N–O in Fig. 129, close against the side of the pattern
rafter while the tongue of the square extends square across the edges
of the two rafters with the outside edge of the tongue of the square
on point O, Fig. 129. Then, make a mark square across the edges
of the two rafters along the outside edge of the tongue of the square
(line O–U in Fig. 129).

This will locate point U on that side of the stuff which is not
close against the pattern rafter. Make a mark across the edge of
the stuff joining points U and P, Fig. 129, and this mark will show
the side cut for that edge of the stuff which will be the back of the
rafter. Having found and marked point U on the edge of the back

of the rafter, you can determine the plumb cut for the side of the rafter, which is not close against the pattern rafter by using the steel square as you did for getting the plumb cuts on the two sides of the pattern rafter. To do this, take away the pattern rafter in Fig. 129 and lay the other rafter flat down on the sawhorses so that the side of the stuff which was close against the pattern rafter in Fig. 129 will be underneath and resting on top of the sawhorses while the other side of the stuff is uppermost and facing you. Point U will then be on the edge of the stuff which is farthest away from you, as shown in Fig. 130.

Now hold the steel square so that the heel is pointing toward you, with the blade at your left, and place the square flat down on the side of the stuff with the 15¹³⁄₁₆-inch division of the outside edge of the tongue of the square on the point U, Fig. 130, and with the

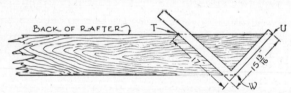

Fig. 130. Steel Square Used for Plumb Cut for Opposite Hand Rafters

17-inch division of the outside edge of the blade of the square on the same edge of the stuff (that edge which will be the top edge or back of the rafter) farther along to the left at point T. With the square in this position, make a mark across the side of the stuff along the outside edge of the tongue of the square from point U to the opposite edge of the stuff at point W. This mark will show the plumb cut for this side of the hip rafter; now, since our first step was tracing the plumb cut for the other side, the plumb cuts and the side cut have all been marked out for the two opposite hand hip rafters. One rafter may be cut along these marks and then used as a pattern for cutting the other. The seat cut and the cuts for the tails at the lower ends of these rafters are the same as for the first hip rafter which was cut. This first hip rafter will be used as a pattern for cutting the lower ends of the other three hip rafters.

**Hip=Jack Rafters.** The common rafters and the hip rafters have been laid out and cut, so the next jobs on which the steel square can be used will be the laying out and the cutting of the valley rafters and jack rafters. In this roof there will be many hip jacks so they will be dealt with now. Figs. 105 and 107 show that in the main roof the ridge is quite short; because of this, many of the rafters will have their upper ends resting against the hip rafters instead of against the ridge board and will therefore be hip jacks. The lower ends of these rafters will be cut the same as the common rafters, but the upper ends will be different because they rest against the hip rafters instead of against the ridge board.

The hip jacks will be of different lengths depending upon their spacing and upon the distance of the center of the rafter from the corner where the two wall plates meet, measured horizontally along the outside face of the wall plate. In this roof the rafters will be spaced 16 inches apart, center to center. The first hip jack from the corner (and therefore the shortest one) will be 16 inches from the corner where the two wall plates meet. The second hip jack will be 32 inches from the corner. The third hip jack will be 48 inches from the corner, and so on. Their lengths will vary according to their distance from the corner. If the rise per foot run of the common rafters in the roof were an even number of inches, such as 8 inches or 9 inches, the length of the shortest jack rafter and also the difference in length between the first and second jack and between the second and third jack, and so on, could be found by looking at the tables on the face of the blade or body of the steel square. In this roof, however, the rise per foot run is $15\frac{13}{16}$ inches, so these lengths must be determined by some other method. The steel square can be used for this purpose (employing the method which is illustrated by Figs. 71 and 72) as follows:

**Determining Lengths of Hip=Jack Rafters.** Take the steel square and, using the outside edge of the blade as a ruler, make a straight line or mark 2 feet long on any fairly smooth surface, such as a large piece of brown paper laid flat on the floor, or even on the dressed surface of the floor itself. By moving the square along, lengthen this line to about 40 inches. Now, hold the square with the heel pointing away from you and the blade on your right and lay it down flat across the mark just made, as shown in Fig. 131, in such a way that the 12-inch division of the outside edge of the tongue of the square and the $15\frac{13}{16}$-inch division of the outside edge of the blade of the square (equal to the rise per foot run of the common rafters) are both on the mark (O–P, Fig. 131). With the square in this position, make a mark (O–N, Fig. 131) along the outside edge of the tongue of the square and extend it beyond the heel of the square about 8 inches.

Keeping the outside edge of the tongue of the square always on this mark, move the square along the mark O–N away from you until the 16-inch division of the outside edge of the tongue is at point O, Fig. 131, where the 12-inch division of the outside edge of the tongue was before, as shown by the dotted lines in Fig. 131. With the square in this position, measure off with a rule the distance O–P between the points where the outside edge of the tongue and

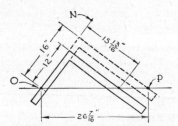

Fig. 131. Finding Length of Hip-Jack Rafters with Steel Square

the outside edge of the blade of the square cross the line O–P, Fig. 131. This distance will be found to be $26\frac{7}{16}$ inches, and this (plus an allowance for the tail of the rafter) will be the length required for the first jack rafters at each of the four corners of the main roof. It will also be the difference in length between any two of the hip-jack rafters which are 16 inches apart on centers.

**Spacing Hip=Jack Rafters.** In the case of the main roof, the horizontal distance in plan view from the outside face of the wall plate on any outside wall to the center line of the ridge board, or to the point where the center line of the ridge board meets the center lines of the hip rafters, is 14 feet and 1 inch, which is 14 times 12 plus 1, or 169 inches. Therefore, there will not be an even number of 16-inch spaces, in plan view, between the outside face of the wall plate at any one of the four corners of the main roof and the point where the center line of the ridge board meets the center lines of the hip rafters. If you try to place a common rafter so that its center line will come at the point where the center line of the ridge board meets the center lines of hip rafters, the spacing of the hip jacks will not work out well; besides, you will get a complicated framing at the ends of the ridge boards, with the hip rafters and the common rafters all meeting at the one point.

It will be better to start at the corners of the main roof where the outside faces of the wall plates meet (*E* and *I*, Fig. 105) and space the hip-jack rafters 16 inches apart from these points so that the longest hip-jack rafters will come several inches away from the points where the center lines of the hip rafters meet the center line of the ridge board, *H* and *F*, Fig. 105. In this way, the length of your shortest hip jack will be $26\frac{7}{16}$, say $26\frac{1}{2}$, inches (see Fig. 131) plus an allowance of 24 inches for the tail which will be the same as for the tail of a common rafter, making a total length of 4 feet, $2\frac{1}{2}$ inches for the shortest hip jacks. The next shortest hip jacks will be $26\frac{1}{2}$ inches longer than the shortest ones and will therefore have a total length of about 6 feet, 5 inches. The next hip jacks will be $26\frac{1}{2}$ inches longer, making a total length of about 8 feet, $7\frac{1}{2}$ inches, and so on for the other hip jacks, each one $26\frac{1}{2}$ inches longer than the next shorter jack rafter.

**Marking for Seat and Tail Cuts of Hip=Jack Rafter.** Pick out a number of pieces of 2- by 6-inch stuff or 2- by 8-inch stuff a little longer than the total lengths mentioned, and dress one edge of each piece, this edge to become the top edge or *back* of each hip-jack rafter. Using a gauge, make a mark straight down the center of the dressed edge of the stuff (the back of the rafter) to serve as a measuring line. With a rule or the edge of the steel square, measure off 24 inches from the left-hand end of the stuff to allow for the tail of the rafter and make a mark on the measuring line at this point. This mark will locate the point where the line of the outside face of the wall plate would intersect the center line of the back of the hip-jack rafter. With a try-square, make a mark square across the back of the rafter at this point (point *A* in Fig. 132); then lay out and make the seat cut and the cuts for the tail of the rafter, using the steel square exactly as was described for the common rafter.

**Marking for Top Cut of Hip=Jack Rafters.** Starting at point *A*, Fig. 132, lay off with a rule to the right along the dressed edge of the stuff the length of the hip-jack rafter, which will be either 2 feet, 2½ inches or two or three or four or more times this distance depending upon the hip-jack rafter with which you are dealing. This will give you a point on the measuring line (point *O*, Fig. 132) which locates where the center line of the back of the hip-jack rafter meets the line of the hip. Having located this point, the next task

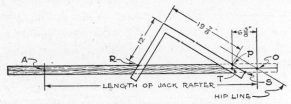

Fig. 132. Laying Out Hip-Jack Rafters by Means of Steel Square

is to lay out the side cut on that end of the stuff which will be the upper end of the hip-jack rafter, after making suitable allowances for the thickness of the main hip rafter. This allowance is necessary because the upper end of the hip jack must rest against the side of the hip rafter; and the amount to be allowed, as illustrated in Fig. 77 at *A*, depends upon the thickness of the hip rafter.

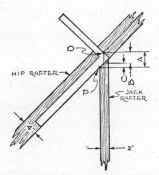

Fig. 133. Using Square with Jack Rafters to Allow for Thickness of Hip Rafter

In the roof frame, the hip rafter is 4 inches thick and the jack rafter is 2 inches thick. Fig. 133 shows these two rafters in plan view. At point *O*, the center line of the back of the hip-jack rafter meets the hip line, which in this plan view is also the center line of the back of the hip rafter. The true distance *A*, Fig. 133, between point *O* and point *P* on one edge of the back of the hip-jack rafter is what you must find in order to make the side cut and plumb cuts for the jack rafter so it will fit against the hip rafter. The distance A in the plan view is made up of two parts, B and C, Fig. 133. Since the hip-jack rafter makes an angle of 45 degrees with the hip rafter in the plan view, the distance C in the plan view, Fig. 133, is equal to one-half of the thickness of the jack rafter. In this case, it is 1 inch because the thickness of the hip-jack rafter is 2 inches. The distance B in plan view, Fig. 133, is equal to the diagonal of a square, the sides of which are one-half the thickness of the hip rafter or in this case 2 inches because the hip rafter is 4 inches thick. To find what distance B is, place the steel square flat down across the dressed edge or back of the hip rafter, as shown in Fig. 133, with the outside edge of the blade exactly along one edge of the back of the rafter and the tongue extending straight across the back of the rafter. Measure off and mark along the outside edge of the tongue a distance equal to one-half the thickness of the hip rafter measured from the heel of the square (point O) and an equal distance from the heel of the square measured along the blade of the square. Each of these distances will be about 2 inches. With a rule measure off the diagonal distance across the square between these two marks; it is approximately $2^{13}/_{16}$ inches. This will be the distance B in Fig. 133.

You now know that distance C in Fig. 133 is 1 inch and this, added to distance B, makes $3^{13}/_{16}$ inches, which is the total for distance A, Fig. 133, *in plan view* (along the run of the hip-jack rafter).

To locate the proper place to make the side cut and the plumb cuts at that end of the stuff which will be the upper end of the hip-jack rafter, it is necessary to know distance A, Fig. 133, not in plan view or along the run of the hip rafter but along the *length* of the rafter. To help in finding this, it is shown in Fig. 134 (where the $15^{13}/_{16}$ represents the rise per foot run of the common or jack rafters) that, for every foot (12 inches) in the run of the common rafters and the jack rafters, the length of the rafter measured along the center line of the back of the rafter will be $19\frac{7}{8}$ inches. Therefore, Fig. 133, the distance A measured along the *length* of the jack rafter will be to the distance A measured along the *run* of the jack rafter ($3^{13}/_{16}$ inches), as $19\frac{7}{8}$ inches is to 12 inches. This problem can be worked out by arithmetic but it can be solved more easily by means of the steel square as follows:

On any fairly smooth surface such as a large piece of brown paper stretched out on the floor or even on the floor itself, mark out a straight line about 2 feet, 6 inches long. Holding the steel square with the heel pointing toward you and with the blade at your right hand, lay it flat down across the line $A-B$ in Fig. 134 in such a way that the 12-inch division of the outside edge of the tongue of the square is on the line at point $A$ and the $15\frac{13}{16}$-inch division of the outside edge of the blade of the square (equal to the rise per foot run of the common or jack rafters) is also on the line at point $B$ as shown in Fig. 134. The distance along the line $A-B$, Fig. 134, between the outside edges of the tongue and blade of the square is $19\frac{7}{8}$, which is the length per foot run for the jack rafters.

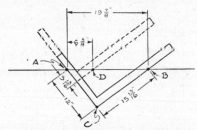

Fig. 134. Using Square with Jack Rafters to
Allow for Thickness of Hip Rafter

Make a mark $A-C$ along the outside edge of the tongue of the square for its full length and move the square along this mark away from you to the left until the $3\frac{13}{16}$-inch division of the outside edge of the tongue of the square is at point $A$, Fig. 134, on the mark where the 12-inch division of the outside edge of the tongue of the square was before. With the square in its new position shown by dotted lines in Fig. 134, measure the distance between point $A$ and point $D$, where the outside edge of the blade of the square in its new position crosses the line $A-B$. This distance, $6\frac{3}{8}$ inches, will be the distance $A$ in Fig. 133 measured along the center line of the back of the hip-jack rafter in its *sloping* position in the roof frame.

Now that the true distance $A$ in Fig. 133 is known, it is possible to lay out the side cut on that end of the stuff which will be the upper end of the hip-jack rafter, making suitable allowance for the thickness of the hip rafter. The center line of that edge of the stuff which will be the back of the hip-jack rafter (line $A-O$, Fig. 132) and the point where the center line of the jack rafter would meet the hip line (point *O*, Fig. 132) have already been marked off. Measure back along the measuring line the distance *A–D*, Fig. 134 (6⅜ inches) and mark point *T*, Fig. 132, on the measuring line. With the steel square or a try-square, make a mark *P–T* straight across the edge of the stuff through point *T* and at right angles with the measuring line. This gives the location of point *P*, Fig. 132, on the edge of the back of the stuff; through this point the side cut must pass.

The method which can be used for marking off the side cut is illustrated in Fig. 135. Here the hip rafter and the jack rafter make an angle of 45 degrees with each other because this is a plan view

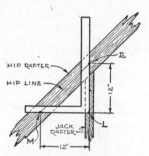

Fig. 135. Use of Steel Square
to Lay Out Side Cut for Jack
Rafters

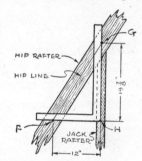

Fig. 136. Use of Steel Square
to Lay Out Side Cut for Jack
Rafters

of the rafters and shows the run of the jack rafter instead of its true length. Therefore, the 12-inch division of the outside edge of the blade of the square, and the 12-inch division of the outside edge of the tongue of the square will both be on the hip line *M–E*, Fig. 135, when the outside edge of the blade of the square lies along the center line of the back of the hip-jack rafter *E–L* in the plan view, Fig. 135.

In Fig. 136, the steel square is shown as it would appear if it were laid down flat on the *sloping* surface of the roof. In this case, if the 12-inch division of the outside edge of the tongue of the square were on the hip line as at *F*, Fig. 136, then the division of the outside edge of the blade of the square which would lie on the hip line as at *G*, Fig. 136, would have to be the same as the length per foot run of the hip-jack rafter which, in the roof with which you are dealing, would be 19⅞ inches instead of 12 inches. In the view shown in Fig. 136, line *F–G*, which might be drawn diagonally across between the two divisions just mentioned on the blade and tongue of the square, lies along the hip line and neither the blade nor the tongue of the square crosses the back of the hip-jack rafter. In Fig. 137 the square is shown swung around and turned over so that the line *O–T* between the 19⅞-inch division of the outside edge of the blade and the 12-inch division of the outside edge of the tongue lies along the center line of the back of the hip-jack rafter instead of along the hip line. In this way, the blade of the square

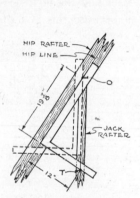

Fig. 137. Use of Steel Square
to Lay Out Side Cut for Jack
Rafters

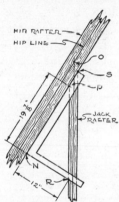

Fig. 138. Use of Steel Square
to Lay Out Side Cut for
Jack Rafters

lies along the hip line in such a position that it would cross the back of the hip-jack rafter at point *O*, Fig. 137, if this rafter were to be extended to the hip line.

In Fig. 138 the square is shown moved along the line of the hip-jack rafter so that the outside edge of the blade of the square will lie along the side of the hip rafter (line *N–P–S*) across the upper end of the hip-jack rafter and will pass through point *P*, thus showing the line of the side cut across the back of the hip-jack rafter, line *P–S*, Fig. 138.

Following the method explained above, hold the square with the heel pointing away from you and with the blade at your right, and lay it flat across that edge of the stuff which will be the top edge or back of the hip-jack rafter (as shown in Fig. 132) in such a

way that the 19⅞-inch division of the outside edge of the blade of
the square will come at point $P$ on the edge of the back of the rafter,
while the 12-inch division of the outside edge of the tongue of the
square rests on the same edge of the back of the rafter farther along
to the left, at point $R$. With the square in this position, a mark
made along the outside edge of the blade of the square will be the
mark for the side cut for the hip-jack rafter. This mark will pass
through point $P$ on one edge of the back of the jack rafter and will
locate point $S$ on the other edge of the back of the hip-jack rafter.
These two points (which will be marked not only on the edges of

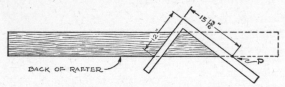

Fig. 139. Laying Out Plumb Cuts for Hip-Jack Rafters by Means
of Steel Square

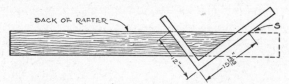

Fig. 140. Laying Out Plumb Cuts for Hip-Jack Rafters by Means
of Steel Square

the back of the hip-jack rafter but also on the two sides of the stuff
on the edges where the sides intersect the top) will serve as starting points for marking out the plumb cuts which have to be made,
one on each side of the hip-jack rafter.

To make the two plumb cuts, apply the steel square to the sides
of the stuff, as shown in Figs. 139 and 140, with the heel of the square
pointing away from that edge of the stuff which will be the top edge
or back of the hip-jack rafter and with the 15¹³⁄₁₆-inch division of
the outside edge of the blade of the square (equal to the rise per foot
run of the common rafters) at point $P$ (Fig. 139) or point $S$ (Fig.
140) while the 12-inch division of the outside edge of the tongue
of the square is also resting on the same edge of the stuff, farther
over to the left. With the square in this position, a mark made
across the side of the stuff along the outside edge of the blade of the

square from point $P$ (Fig. 139) or point $S$ (Fig. 140) to the opposite edge of the stuff (the lower edge of the hip-jack rafter) will give you the plumb cut for each side of the hip-jack rafter. You will then be able to make the cuts for the upper end of one hip jack; use this one as a pattern for cutting all of the other hip-jack rafters which will be the same as this one except for the difference in length.

**Valley Rafters.** Valley rafters which extend from the wall plate all the way up to the ridge would be the same as the hip rafters but, in your roof, there are no valley rafters of this kind. The other valley rafters will be shorter and will have their upper ends resting against the hip rafters. There are three valley rafters of this sort in your roof, marked $A$, $B$, and $C$ in the roof plan, Fig. 105. They will be of different lengths. There is another valley rafter in your roof, marked $D$ in Fig. 105, which is shorter than any of the others and has its upper end resting against valley rafter $B$. In the plan view, all of these valley rafters will be at right angles to the longer valley rafter, such as rafter $B$, or to the hip rafters against which their upper ends rest, just as the common rafters are at right angles to the ridge board in plan view. Because the sides of both the valley and the hip rafters are vertical—that is, *plumb*—you will have at the upper end of each valley rafter a plumb cut similar to the plumb cut at the upper end of the common rafter, except that the bevel will be different. The bevel will be different because the run of each of the valley rafters is 17 inches for every foot of run of the common rafters. However, the side cut for the valley rafters will be made square across the thickness of the valley rafter, just like the side cut at the upper end of a common rafter where it rests against the ridge board.

**Determining Lengths of Valley Rafters.** Of the four valley rafters marked $A$, $B$, $C$, and $D$ in Fig. 105, the one marked $C$ is the only one which has a *tail*. The lower ends of each of the other valley rafters will terminate at the wall plate. Valley rafters $A$, $C$, and $D$ have the ridge boards of the lower roofs over the ells at their upper ends, which means that the lengths of these valley rafters can be found directly from the run and rise of the common rafters in the low roofs over the ells, as was done in the case of the hip rafters. Valley rafter $B$, however, has to be carried up beyond the ridge of the ell roof so that its upper end can rest against the hip rafter

(*F–G*, Fig. 105). For this reason, the length of the valley rafter *B* will have to be found by using the distance between the lower end of the valley rafter *B* in Fig. 105 and the lower end of the hip rafter *F–G*, together with a rise of 15¹³⁄₁₆ inches for each 2 feet in this distance or, in other words, for each foot in half the distance.

To find the length of valley rafter *B*, then, measure to scale the distance *P–R* in Fig. 108 and find it to be 15 feet, 4 inches, one-half of which will be 7 feet, 8 inches. Now take the steel square and lay it flat down on a smooth surface such as a large piece of brown paper stretched out on the floor or on the floor itself, with the back of the square up so that you will be looking at the side of the square which has the outside edges divided off into inches and twelfths of an inch. Draw a line along the outside edge of the blade which will be placed nearest to you and another line along the outside edge of the tongue of the square which will be at your right and

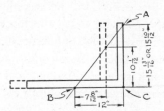

Fig. 141. Using Square to Find Rise
for Valley Rafter

pointing away from you. See Fig. 141. You will find the 12-inch division of the outside edge of the blade of the square, *C–B*, Fig. 141 (equal to one foot of the run of a common rafter of the ell roof) and make a mark there, *B*. You will also find the 15¹⁰⁄₁₂-inch division of the outside edge of the tongue of the square (equal to 15¹³⁄₁₆ inches, which is the rise per foot run of the common rafters in the ell roof) and make a mark there (*A*, Fig. 141).

Now lift the square and join the points *A* and *B* with the line *A–B* representing the slope of the ell roof, the line *C–B* representing the run of the rafters, and the line *A–C* representing the rise of the rafters. Lay the square down again with the outside edge of the blade of the square still along the line *C–B*, Fig. 141, but place the square so that the 7⁸⁄₁₂-inch division of this edge of the square (equal in inches to one-half the distance *P–R*, Fig. 108, in feet) is at the mark $B$, Fig. 141, as shown by the dotted outline of the steel square in Fig. 141. With the square in this position, the outside edge of the tongue of the square will cross the line $A$–$B$ at the $10^1/_{12}$-inch division of the outside edge of the tongue of the square, corresponding, at a scale of 1 inch equals 1 foot, to a distance of 10 feet and 1 inch which is the total rise of valley rafter $B$, Fig. 105. The total run of valley rafter $B$ will be equal to 17 inches or $1^5/_{12}$ feet for each foot in one-half the distance $P$–$R$, Fig. 108. Since the distance $P$–$R$, Fig. 108, is 15 feet and 4 inches, one-half of this dimension will be $7^8/_{12}$ feet; and the total run of the valley rafter $B$, Fig. 105, will be $1^5/_{12}$ times $7^8/_{12}$ feet, or $^{17}/_{12} \times 7^8/_{12}$ feet.

To find this, put the steel square down flat as before, draw lines along the two outside edges, take the 12-inch division of the outside edge of the tongue and the 17-inch division of the outside edge of the blade, and mark them ($A$ and $B$, Fig. 142). Raise the square,

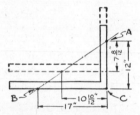

Fig. 142. Using Square to Find Run for Valley Rafter

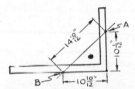

Fig. 143. Using Square to Find Length for Valley Rafter

and draw a line diagonally across between these two points, as shown in Fig. 142 by line $A$–$B$. Put the square down flat again with the outside edge of the tongue along the same line as before (line $A$–$C$, Fig. 142) but with the $7^8/_{12}$-inch division of this edge, equal to one-half the distance $P$–$R$, Fig. 108, at point $A$ in Fig. 142. With the square in this position, the outside edge of the blade of the square will cross the line $A$–$B$, Fig. 142, at the $10^{10}/_{12}$-inch division of this edge of the blade, which equals (to a scale of 1 inch to 1 foot) the distance 10 feet and 10 inches; this distance is the total run of the valley rafter $B$, Fig. 105. Lay the steel square down again and mark (Fig. 143) along the outside edges of the tongue and the blade. Find the $10^1/_{12}$-inch division of the outside edge of the tongue (equal to the total rise of valley rafter $B$, Fig. 105, at a scale of 1 inch equals 1 foot). Find the $10^{10}/_{12}$-inch division of the outside edge of the blade (equal to the total run of valley rafter $B$, Fig. 105, to scale). Measure with a rule the distance ($A$–$B$, Fig. 143) diagonally across the square between these two points and find it to be $14^{10}/_{12}$ inches which represents (to scale) a distance of 14 feet and 10 inches, the length of valley rafter $B$, Fig. 105.

**Marking Plumb Cuts for Upper End of Valley Rafters.** Select a piece of 4- by 10-inch stuff about 15 feet long to be sure it is long enough and dress one edge which will be the top edge or back of the valley rafter $B$, Fig. 105. With a square, make a mark square across the dressed edge of the stuff about 1 foot from one end, which will be the lower end of the valley rafter as shown in Fig. 144, and mark

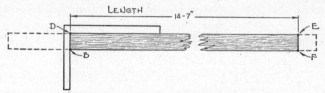

Fig. 144.  Laying Out Valley Rafter with Steel Square

well the ends of this line at the sides of the stuff (points $B$ and $D$, Fig. 144). These marks show where the outside face of the wall plate (if it were extended up to the top of the rafter) would intersect or meet the back of the rafter.

From this mark measure along the dressed edge of the stuff a distance equal to the length of valley rafter $B$, Fig. 105, less 3 inches or one and one-half times half the thickness of the hip rafter against which the valley rafter rests. This distance in your case will be 14 feet and 10 inches, less 3 inches, which equals 14 feet and 7 inches. This, when measured off along the dressed edge of the stuff, will locate the position of the plumb cut at the upper end of valley rafter B. A mark made square across the dressed edge of the stuff by means of the steel square at this point ($E$–$F$, Fig. 144) will locate the top of the plumb cut on each side of the rafter at points $E$ and $F$, Fig. 144.

To make the plumb cut, place the steel square flat against the side of the stuff, holding it so the heel of the square points away from that edge of the stuff, which will be the back of the rafter, as shown at the right-hand end of Fig. 145. Place the $15^{13}/_{16}$-inch division of the outside edge of the tongue of the square at the mark $F$

on the dressed edge of the stuff, which represents the top of the plumb cut; and place the 17-inch division of the outside edge of the blade of the square on the same dressed edge of the stuff farther along to the left, at point *G*, Fig. 145. A mark made across the side of the stuff along the outside edge of the tongue of the square will show the plumb cut to be made as on line *F–H*, Fig. 145.

**Making the Seat Cut for Valley Rafters.** To make the seat cut, lay the stuff down flat with the dressed edge, or back, away from you. Place the steel square against the side of the stuff with the heel of the square also pointing away from you and with the tongue at your left; see *B–K–L*, Fig. 145. Place the 15¹³⁄₁₆-inch division of the outside edge of the tongue of the square at mark *B*, which represents

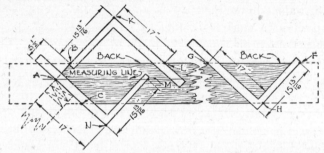

Fig. 145.    Laying Out Plumb Cut and Seat Cut for Valley Rafter with Steel Square

the intersection of the line of the outside face of the wall plate with the back of the rafter, while the 17-inch division of the outside edge of the blade of the square is at point *L* on the same dressed edge of the stuff, farther along to the right. A mark made along the line of the outside edge of the tongue of the square, using a rule placed against this edge to extend it across the side of the stuff, will show the vertical portion of the seat cut (line *A–B*, Fig. 145). When the valley rafter is raised up into its proper sloping position in the roof frame, this line will be vertical or plumb.

To mark out and make the horizontal portion of the seat cut so the lower end of the valley rafter will rest on top of the wall plate, first find out where this part of the seat cut will be. To locate it, measure off from point *B*, Fig. 145, along the line *B–A* and away from the dressed edge of the stuff the distance *B–A*, 5½ inches, which will be equal to the similar distance on the common rafters

(*A–B,* Fig. 119A). This will give point *A,* which marks the intersection of the outside upper corner of the wall plate with the side
of the valley rafter or rather with the center of the valley rafter.
Using a gauge, mark off on the side of the stuff a line (*A–M,* Fig. 145)
passing through point *A,* parallel to the dressed edge of the stuff
and about 28 inches long, which will be our measuring line.

Now take the square and hold it with the heel pointing toward
you and away from the dressed edge of the stuff (as at *A–N–M,*
Fig. 145) and with the tongue at the right. Lay the square down
on the side of the stuff with the 17-inch division of the outside edge
of the blade on point *A,* Fig. 145, and with the 15¹³⁄₁₆-inch division
of the outside edge of the tongue of the square on the measuring
line farther along to the right, as shown in Fig. 145. With the square
in this position, a mark *A–C,* Fig. 145, made across the side of the
stuff along the outside edge of the blade of the square from point *A*
to that edge of the stuff which will be the lower edge of the valley
rafter will show the horizontal portion of the seat cut for the valley
rafter (*A–C,* Fig. 145). Valley rafters *A* and *D,* Fig. 105, will be
similar to valley rafter *B* except for their length; and valley rafter *C,*
Fig. 105, will be similar also except that it will have a tail like those
at the lower ends of the hip rafters.

**Laying Out and Cutting Valley=Jack Rafters.** To lay out and
cut the valley-jack rafters which bear against the ridge board at
their upper end and against the valley rafters at the lower end, first
of all find the run of each of them by measurement from the roof
plan (Fig. 105). Assuming that they are 16 inches apart on centers
and that they are spaced regularly, starting at the point where the
valley rafter and the ridge board meet, the run of the shortest valley-
jack rafters will be 16 inches; the run for the next longest ones will
be twice as much, or 32 inches; the run for the next longest ones
will be three times 16 inches, or 48 inches, and so on. The lengths
of the valley jacks will be in the same proportion—that is, the length
for the shortest valley jack will be found and this length multiplied
by 2 for the next longest one and by 3 for the next longest and by
4 for the next longest, and so forth.

To find the length for the shortest valley jack, do as you would
for finding the length of the shortest hip-jack rafter. Lay the square
down flat on a smooth surface with the outside edge of the tongue

toward you; see Fig. 146. Draw a line along the outside edge of the tongue and make a mark at the 12-inch division of this edge of the square (point $A$, Fig. 146). Draw a line along the outside edge of the blade and make a mark at the 15$\frac{13}{16}$-inch division (equal to the rise per foot run of the roof) of this edge of the square (point $B$, Fig. 146). Lift the square and draw a line which joins the two marks and extends farther at each end, line $A$–$B$–$C$. Extend the line which was drawn along the outside edge of the tongue of the square ($A$–$D$, Fig. 146) and lay the square down again with the outside edge of the tongue on this line but with the 16-inch division at the mark $A$, where the 12-inch division of this edge of the tongue was before, as shown by the dotted lines in Fig. 146.

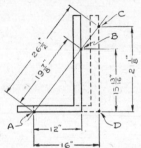

Fig. 146. Finding Length of Valley
Jack with Steel Square

Now see where the outside edge of the blade of the square in its new position (shown dotted in Fig. 146) crosses the diagonal line $A$–$B$–$C$, Fig. 146, and mark this point, as at $C$. Then by measuring the distance between the two points $A$ and $C$, Fig. 146, you will have the length of one of your valley jacks or a hip-jack for each 16 inches of run and the length of the shortest valley jack. You can get the lengths of the other valley jacks from this as previously explained. Mark out and make the plumb cuts at the upper ends of the valley jacks in the same way as for the upper ends of the common rafters, allowing for the thickness of the ridge board. Mark out and make the side cut and plumb cuts at the lower ends of the valley jacks where they bear against the valley rafters in the same way as for the upper ends of the hip jacks where they bear against the hip rafters, making proper allowance for the thickness of the valley rafter.

**Laying Out and Cutting Cripple=Jack Rafters.** In your roof there are a number of cripple-jack rafters which have their upper ends bearing against hip rafters and their lower ends bearing against valley rafters without touching either the ridge board or the wall plate. To 'ay out and cut these cripple jacks, you will have to find the run for each rafter by measurement from the roof plan, Fig. 105, in feet and inches. Take a cripple jack for which the run is 7 feet and 7 inches. This run can be reduced for use with the steel square to a scale of 1 inch equals 1 foot, so that 1 inch on the edge of the square equals 1 foot of run, and $^1/_{12}$ of an inch on the edge of the square equals 1 inch of run. Then the 7 feet, 7 inches becomes $7^7/_{12}$ inches. All of the common and jack rafters in your roof have

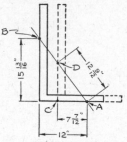

Fig. 147. Finding Length of
Cripple-Jack Rafter with Steel
Square

a rise per foot run of $15^{13}/_{16}$ inches. Lay the square down flat on a smooth surface with the blade extending away from you and with the back facing up, as shown in Fig. 147, so that you are looking at the back of the square. Draw lines along the outside edges of the blade and the tongue, marking the 12-inch division along the edge of the tongue at $A$, and $15^{13}/_{16}$ inches *by measurement with a rule* along the edge of the blade at $B$. Now lift the square and draw a line $A$–$B$, Fig. 147, connecting these two points.

Lay the square down again with the outside edge of the tongue along the same line as before (line $A$–$C$) but with the $7^7/_{12}$-inch division (representing the run of the cripple jack) at the point $A$, Fig. 147, where the 12-inch division was before, as shown by the dotted lines in Fig. 147. With the square in this position, note that the outside edge of the blade crosses the line $A$–$B$, Fig. 147, at point $D$ and that the distance $A$–$D$ is by measurement $12^9/_{12}$ inches.

You know now that the length of this cripple jack is 12 feet and 9 inches.

Now take a piece of stuff a little longer than the length just mentioned; dress one edge, which will be the top edge or back of the cripple jack, and mark out a measuring line down the middle of the dressed edge. On this measuring line, mark two points near the two ends of the piece of stuff, the distance between them being equal to the length of the cripple jack. One of these points will mark the intersection of the measuring line with the center of the hip rafter at the upper end of the cripple jack, and the other the intersection of the measuring line with the center of the valley rafter at the lower end of the cripple jack. With these two points established, mark off and make the side cut and plumb cuts for the upper end of the cripple jack, which will bear against the hip rafter in the same way as was described for the hip-jack rafters, allowing for the thickness of the hip rafter. Also mark off and make the side cut and plumb cuts for the lower end of the cripple jack, which will bear against the valley rafter in the same way as was described for the valley-jack rafters.

**Dormer Window.** The dormer window over the stair landing which is shown on the north elevation in Fig. 106 is what is known as an octagon type of dormer. However, it is not a *true* octagon dormer in plan, since the front wall where the windows are located has a length out-to-out of wall plates from one corner of the so-called octagon to the other corner of 5 feet and 6 inches, while the other two sides which return at a 45-degree angle in plan are only 1 foot, 9 inches long; see Fig. 148. (You arrive at the dimension of 1 foot, 9 inches by taking your steel square and measuring diagonally across with a rule from the 15-inch division on the outside edge of the blade to the 15-inch division on the outside edge of the tongue. This distance is a little more than 1 foot, 9 inches.)

The distance across the so-called half-octagonal plan formed by the outside edges of the wall plates at the front of the dormer window from one side to the other is 1 foot and 3 inches plus 5 feet and 6 inches plus 1 foot and 3 inches or a total of 8 feet; the distance from the outside of the wall plate on each side of the dormer to the ridge line of the dormer roof is one-half of this, which is 4 feet or 48 inches. See Fig. 148. Since this is a roof of equal pitch, 48 inches is also the distance from the outside edge of the wall plate at the front wall of the dormer straight back to the point where the hip lines will meet the line of the ridge in plan. One-half of the length of the front wall plate is 2 feet and 9 inches, or 33 inches; see Fig. 148. These distances are greater than the length of the blade of the steel square so, to find the run of the hip rafters of the roof over the octagonal type dormer, divide each of these distances by 2 and find *half* the length of the run by taking the 16½-inch division (half of 33) on the outside edge of the tongue of a steel square having an 18-inch tongue and by measuring diagonally across to the end of the outside edge of the 24-inch blade (half of 48). You will find one-half of the

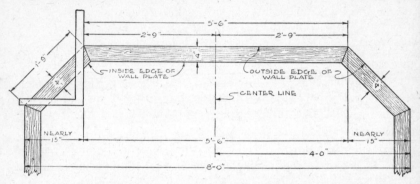

Fig. 148. Layout of Wall Plates for Octagon Type Dormer Window

run of the hip rafter to be 29 inches; the whole run will be twice this distance, or 58 inches.

The roof of the octagon dormer will have a ⅜ pitch, so the total rise will be 3 feet or 36 inches—9 inches per foot run of the common rafters. Then the rise for a run of 2 feet will be 18 inches and this is also the rise for a run of 29 inches (which is half of the total run) of the hip rafter. This is so because the total rise for the hip rafter is the same as the total rise for the common rafters and therefore the rise for one-half the run of the hip rafter will be the same as the rise for one-half of the run of the common rafters.

To find one-quarter of the length of the hip rafter, take one-half of 29 inches, or 14½ inches (which is one-quarter of the run of the hip rafter), and 9 inches (which is one-quarter of the rise of the hip rafter). You will find the 14½-inch division on the outside edge of the blade of the square and the 9-inch division on the out-

side edge of the tongue, as shown in Fig. 149, and you will measure
the diagonal distance between these two points across the square,
distance *A–B*, Fig. 149. The distance is 17 inches, and 4 times 17
will be 68 inches, which is the length of the hip rafter. All hip rafters
are the same length.

**Determining Seat, Side, and Plumb Cuts for Hip Rafters on
Dormer.** Select a piece of 3- by 4-inch stuff a little longer than 68
inches and lay out the hip rafter. At the lower end there will be a
seat cut similar to those for the hip and valley rafters in the main
roof but without a tail. At the upper end there will be a side cut

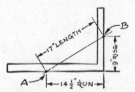

Fig. 149. Finding One-Fourth of Length
of Hip Rafter for Octagon Dormer

to fit the hip rafter against the ridge board and plumb cuts similar
to those for the hip rafters of the main roof but with a different
bevel. There are four hip rafters but they are in pairs.

First find the cuts for one of the pairs of hip rafters nearest to
the front of the dormer. You know that the distance from the peak
of the roof to the outside face of the wall plate of the front wall of
the dormer is 4 feet, or 48 inches in plan view (see Fig. 150), and that
half the length of the front wall of the dormer measured along the
outside edge of the wall plate is 2 feet, 9 inches, or 33 inches. The
run of the hip rafter is 58 inches and the length of the hip rafter is
68 inches. Your first step will be to draw the line *B–C*, Fig. 150,
at right angles to the line which represents the run of the hip rafter
*A–B*, and to find its length from the lower end of the hip rafter
(point *B*, Fig. 150) to its intersection with the line of the ridge
board (extended) at point *C*, Fig. 150. This length will bear the same
relation to the length of the run (58 inches) as 33 bears to 48. Since
the lengths are all longer than the blade and the tongue of the square,
you will have to work with one-quarter of each dimension and then
multiply the result by 4. One-quarter of 48 is 12, one-quarter of 33
is $8\frac{1}{4}$, and one-quarter of 58 is $14\frac{1}{2}$.

Lay the square down flat on a smooth surface with the face uppermost; see Fig. 151. Draw lines along the outside edges of the blade and tongue and mark the 12-inch division of the blade (*B*, Fig. 151) and the 8¼-inch division of the tongue, *A*, Fig. 151. Extend the line *A–C* drawn along the edge of the tongue, then lift the square and draw a long line through the two marks *A* and *B*. Lay the square down again with the edge of the tongue along the

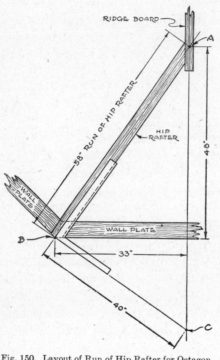

Fig. 150. Layout of Run of Hip Rafter for Octagon Type Dormer

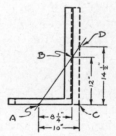

Fig. 151. Using Steel Square to Find Length of Line B–C, Fig. 150

same line as before (*A–C*) but with the square moved along to the right (as shown by the dotted lines in Fig. 151) far enough so the 14½-inch division of the outside edge of the blade of the square is on the long diagonal line (*A–B* at point *D*).

The 10-inch division of the outside edge of the tongue of the square is now at mark *A*, which was 8¼ inches from the heel of the square before the square was moved. From this you find that 10 inches is one-quarter of the distance *B–C*, Fig. 150, which you are trying to find; and that this distance is 40 inches—4 times 10 inches.  Use this distance together with the length of the rafter in laying out the bevel for the side cut at the upper end of the hip rafter, but, as these distances are both longer than the blade or tongue of the square, use the 10-inch dimension together wit': one-quarter of the length, which is 68 inches divided by 4, or 17 inches, as shown on line *A–B* in Fig. 152.

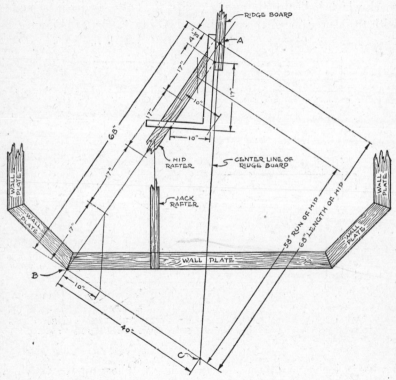

Fig. 152.  Developed Plan of Dormer Showing True Length and Side Cut for Hip Rafter

In other words, in Fig. 150 it is evident that the true bevel for fitting the hip rafter against the ridge board is not the angle *B–A–C* since this is a plan view and shows everything foreshortened.  If you can imagine or visualize the hip rafter swung up into its true sloping position in the roof, you can think of it as being swung up about the line *B–C* as a hinge.  Then length *B–A* would increase to 68 inches, and length *C–A* would increase as shown in Fig. 152, but distance *B–C* would not increase. It would remain 40 inches. Then the true bevel is as 68 is to 40 or (dividing each by 4) as 17 is to 10.

To lay out and make the side cut, mark a measuring line along the center of the dressed edge of the stuff, which will be the back of the hip rafter and make two marks *A* and *B*, Fig. 153A, 68 inches apart on this line to show the length of the hip rafter. From that mark which indicates the upper end of the rafter or its intersection with the center line of the ridge board (point *A*, Fig. 153A), measure

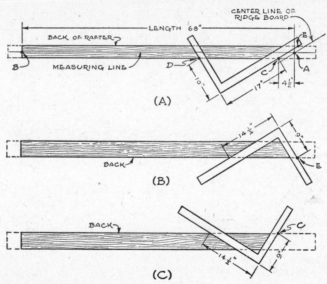

Fig. 153. (A) Laying Out Side Cut at Upper End of Hip Rafter with Steel Square, (B) and (C) Laying Out Plumb Cut at Upper End of Hip Rafter with Steel Square

back 4½ inches along the measuring line to allow for the thickness of the ridge board and the hip rafter and at this point make a mark square across the back of the hip rafter to get point *C*, Fig. 153A, on the edge of the back. Now hold the steel square with the heel pointing toward you and with the blade at your right and place it down across the dressed edge of the stuff with the 17-inch division of the outside edge of the blade on point *C*, Fig. 153A, and the 10-inch division of the outside edge of the tongue of the square on the same edge of the stuff farther along to the left at point *D* in Fig. 153A. With the square in this position, a mark made across the dressed edge of the stuff (the back of the hip rafter) along the outside edge of the blade of the square will show the side cut to be made to fit the hip rafter against the ridge board. The points $C$ and $E$,

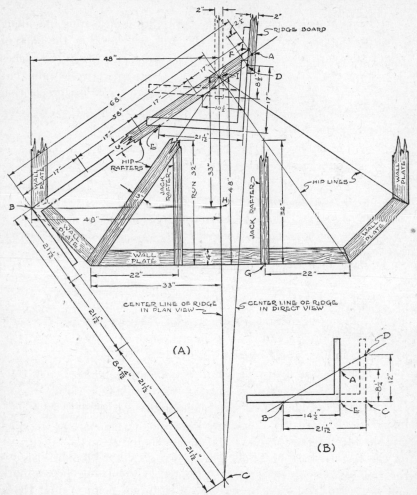

Fig. 154. (A) Developed Plan of Dormer, Using Steel Square to Determine Side Cut for Hip Rafter, (B) Finding Length of Line B–C in (A)

Fig. 153A, where the mark which indicates the side cut meets the two edges of the back of the rafter, show the places from which the two plumb cuts on the sides of the rafter meet the back of the rafter; see Figs. 153B and 153C.

The length of the other pair of hip rafters for the octagon dormer is also 68 inches and the run is 58 inches, the same as for the first pair of hips. For the hips now to be considered, the length of the line $B$–$C$, Fig. 154A (from the lower end of the hip rafter, point $B$, at right angles to its run and extending over to the point $C$, where it intersects the line of the ridge board, extended) bears the same relation to 48 inches ($B$–$H$) as the hip rafter run of 58 inches ($B$–$I$) bears to 33 inches (distance in plan view from ridge to a line extended across from $B$, distance $I$–$H$, Fig. 154A). Divide each of these dimensions by 4 and, by means of the steel square, you will find by using the following procedure that one-quarter of the length of the line $B$–$C$ is $21^1/_{12}$ inches. Lay the square flat down on its side on a big piece of paper, with the back up so that you are looking at the back of the square as shown in Fig. 154B. This is because the outside edges of the back of the square usually shows twelfths of inches. Find the $8\frac{1}{4}$-inch division (one-quarter of 33) on the outside edge of the tongue of the square, point $A$ in Fig. 154B, and the $14\frac{1}{2}$-inch division (one-quarter of 58) on the outside edge of the blade, point $B$ in Fig. 154B, and make marks on the paper at these two points. Also draw a line along the outside edge of the blade of the square ($B$–$E$ in Fig. 154B) and extend this line about 7 inches or more as shown in Fig. 154B by the line $B$–$E$–$C$. Now lift the square and draw a line through the points $B$ and $A$ as shown in Fig. 154B by the line $B$–$A$–$D$. Now put the square down again on the paper just as it was before and then move the square along to the right as shown by the dotted lines in Fig. 154B (keeping the outside edge of the blade of the square always on the line $B$–$E$–$C$), until the 12-inch division (one-quarter of 48) on the outside edge of the tongue of the square, point $D$ in Fig. 154B, is on the line $B$–$A$–$D$ as shown at point $D$. Now, make a mark at the new position of the heel of the square, point $C$ in Fig. 154B, and read off on the outside edge of the blade of the square the distance $B$–$C$, which will be found to be $21^1/_{12}$ inches.

You will use this dimension of $21^1/_{12}$ inches together with one-quarter of the length of the hip rafter, 68 inches, in laying out the bevel for the side cut at the upper end of the hip rafter. One-quarter of the hip rafter length is 17 inches, the same as before.

Therefore, to lay out the side cut to fit this hip rafter against the ridge board, measure back along the measuring line in the middle of the dressed edge of the stuff a distance of $2\frac{1}{2}$ inches from the point $A$, Fig. 154A (in which it intersects the center line of the ridge) to allow for the ridge board thickness and the rafter thickness; make a mark at $D$ square across the dressed edge of the stuff at this point. With this end of the piece of stuff at your right hand and the dressed edge (which will be the back of the hip rafter) uppermost, hold the square with the heel pointing toward you and the tongue at your right and lay it down flat across the dressed edge of the stuff so that the 17-inch division of the outside edge of the tongue of a square with an 18-inch tongue is at point $D$, Fig. 154A, on the edge of the back of the rafter nearest to you, and so that the $21\frac{1}{12}$-inch division of the outside edge of the blade of the square is on the same edge of the back of the rafter farther along to the left at $E$, Fig. 154A. With the square in this position, a mark $D$–$F$, Fig. 154A, made across the dressed edge of the stuff along the outside edge of the tongue of the square will show the side cut to be made to fit this hip rafter against the ridge board. You would have to have a square with a tongue 18 inches long and even then the outside edge of the tongue of the square would not be long enough to cross the edge of the stuff. See Fig. 154A at $D$. For this reason it will be better to use $8\frac{1}{2}$ inches (half of 17) on the tongue and $10\frac{1}{2}$ inches (half of 21) on the blade as shown by the dotted outline of the square at $D$ in Fig. 154A.

You can mark out the plumb cuts or top cuts for all four of the hip rafters on the sides of the rafters with the help of the steel square, as shown in Fig. 55, using 9 inches (one-quarter of the rise) on the edge of the tongue and $14\frac{1}{2}$ inches (one-quarter of the run of 58 inches) on the edge of the blade, as shown in Fig. 153B and 153C.

Since the four hip rafters theoretically have all their upper ends meeting at the same point at the center of the ridge board, the two pairs of rafters will interfere with each other. Cut and put up the pair of hip rafters farthest from the front of the dormer first and then trim the upper ends of the front pair of hip rafters to fit into the angles between the other pair of hip rafters and the sides of the ridge board. You will make the side cuts for these rafters as previously described and then trim the upper ends so that they will be symmetrical about a line drawn down the center of the dressed edge of the stuff; they will then fit into place. This can be done because your dormer window has a front which is not a true octagon in plan.

**Determining Side Cut of Jack Rafters on Dormer.** To make the side cut on the jack rafters of the octagon type dormer to fit them against the hip rafter, you will find the dimensions to take on the blade and tongue of the square by referring to Figs. 154A and 155. Distance *G–H*, Fig. 155, is the run of the jack rafter while distance *G–K* is the length of the jack, 40 inches. Distance *G–L* measured from the foot of the jack rafter and at right angles to it over to where *G–L* intersects the hip line (*H–L* or *K–L*, Fig. 155) is 22 inches. You will have to use half of these dimensions, or 20 inches and 11 inches. Now take a piece of 2 by 4 stuff about 4 feet long and dress one edge, which will be the top edge or back of the rafter. Draw a line down the middle of this dressed edge to be your

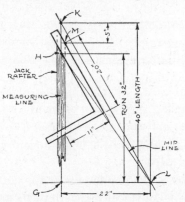

Fig. 155. Use of Square for Side Cut for
Jack Rafter of Dormer

measuring line. Now, since this is a roof of ⅜ pitch, the *length per foot run* for all the common and jack rafters will be 15 inches; see pages 35 and 61. Mark two points 40 inches apart to indicate the length of the rafter ($^{15}/_{12}$ of the run of 32 inches = 40 inches), and place the piece of stuff so that you are looking at the dressed edge or back with the upper end at your right (as shown in Fig. 156). From the point *K* (marking the upper end of the jack rafter), measure back to the left a distance of 5 inches to allow for the thickness of the hip rafter and the jack rafter, and then mark square across the dressed edge of the stuff at this place and find point *M* on the edge of the back of the rafter nearest to you. Hold the square with the heel pointing toward you, with the blade to your right, and lay it down flat across the dressed edge of the stuff with the 20-inch division of the outside edge of the blade at the mark M, and the 11-inch division of the outside edge of the tongue on the same edge of the stuff farther along to the left, at point O, Fig. 156. With the square in this position, a mark M–N made across the dressed edge of the stuff along the outside edge of the blade of the square will show the side cut to be made to fit the jack rafter against the hip rafter.

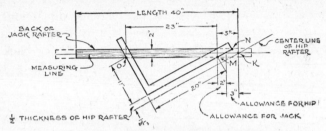

Fig. 156. Laying Out Jack Rafter of Dormer with Steel Square

For the plumb cuts, see Fig. 77, pages 67 and 68, but use 9 inches on the tongue of the square instead of 8 inches because your roof has a ⅜ pitch and the *rise per foot run* is 9 inches.

**Conclusion.** This illustrative problem explains some of the ways in which the steel square can be of use in laying out and cutting the members of the wood framing for a house.

The first four chapters illustrate and explain many of the applications of this almost indispensable instrument, but there are also other useful and interesting ways in which it can be employed to solve the problems which a carpenter may encounter in his work.

To anyone who has made himself familiar with the steel square by a careful study of these chapters, many such applications of the instrument will suggest themselves.

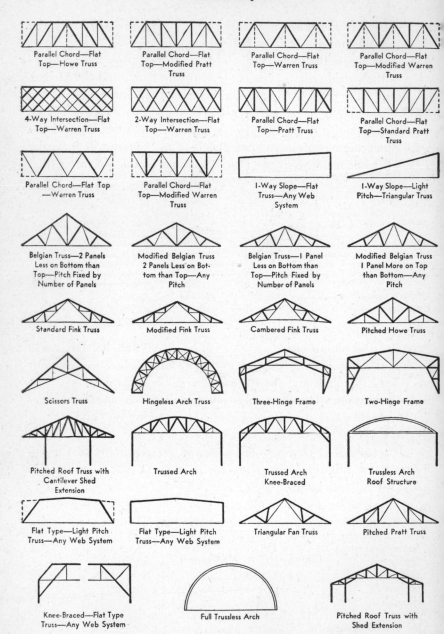

Parallel Chord—Flat Top—Howe Truss

Parallel Chord—Flat Top—Modified Pratt Truss

Parallel Chord—Flat Top—Warren Truss

Parallel Chord—Flat Top—Modified Warren Truss

4-Way Intersection—Flat Top—Warren Truss

2-Way Intersection—Flat Top—Warren Truss

Parallel Chord—Flat Top—Pratt Truss

Parallel Chord—Flat Top—Standard Pratt Truss

Parallel Chord—Flat Top—Warren Truss

Parallel Chord—Flat Top—Modified Warren Truss

1-Way Slope—Flat Truss—Any Web System

1-Way Slope—Light Pitch—Triangular Truss

Belgian Truss—2 Panels Less on Bottom than Top—Pitch Fixed by Number of Panels

Modified Belgian Truss 2 Panels Less on Bottom than Top—Any Pitch

Belgian Truss—1 Panel Less on Bottom than Top—Pitch Fixed by Number of Panels

Modified Belgian Truss 1 Panel More on Top than Bottom—Any Pitch

Standard Fink Truss

Modified Fink Truss

Cambered Fink Truss

Pitched Howe Truss

Scissors Truss

Hingeless Arch Truss

Three-Hinge Frame

Two-Hinge Frame

Pitched Roof Truss with Cantilever Shed Extension

Trussed Arch

Trussed Arch Knee-Braced

Trussless Arch Roof Structure

Flat Type—Light Pitch Truss—Any Web System

Flat Type—Light Pitch Truss—Any Web System

Triangular Fan Truss

Pitched Pratt Truss

Knee-Braced—Flat Type Truss—Any Web System

Full Trussless Arch

Pitched Roof Truss with Shed Extension

Into the skeleton structure of buildings go all types of trusses, many requiring the application of that indispensable tool—the steel square.

# INDEX